Relativity and Cosmology

Second Edition

William J. Kaufmann, III

Jet Propulsion Laboratory
California Institute of Technology

Harper & Row, Publishers
New York Hagerstown San Francisco London

To my daughter
Kristine Nicole Kaufmann
with love

Cover Photo: Edge-on view of the spiral galaxy. NGC 4594 (also know as M 104) in Virgo. (Hale Observatories)

Sponsoring Editor: Wayne E. Schotanus
Project Editor: Eleanor Castellano
Designer: Andrea C. Goodman
Production Supervisor: Stefania J. Taflinska
Compositor: Maryland Linotype Composition Co., Inc.
Printer and Binder: The Maple Press Company
Art Studio: J & R Technical Services, Inc.

RELATIVITY AND COSMOLOGY, Second Edition

Library of Congress Cataloging in Publication Data

Kaufmann, William J.
Relativity and cosmology.

Includes index.
1. Cosmology. 2. General relativity (Physics)
I. Title.
QB981.K3 1977 523 76-41868
ISBN 0-06-043572-0

Index

76 77 78 79 80 9 8 7 6 5 4 3 2 1

Contents

Preface

Astronomy has the dual distinction of being both the oldest science and one of the most fascinating fields of modern research. In the mid-twentieth century, many scientists both young and old turned their attention to a variety of problems of astronomical and astrophysical interest. The result of these efforts is that we seem to be on the brink of a number of fundamental changes in our thinking about the universe as a whole. In particular, it now appears that the general theory of relativity is extremely relevant to any in-depth understanding of physical reality.

One of the unfortunate by-products of the current rapid advance of astronomy and astrophysics is that it is often difficult to keep up with what is going on. Indeed, hardly a month passes without the announcement of some new discovery of basic importance. Whereas the professional astronomer simply must work a little harder to stay informed, the lay public or the casual student of astronomy is often left out in the cold.

This book has been designed specifically to bring to the lay public, the amateur astronomer, and the student of astronomy many of the more exotic and fascinating aspects of current research, particularly general relativity. In this regard, this book will help bridge the gap between current journals and the typical textbooks used in elementary, college-level astronomy courses.

I am deeply grateful to Drs. John A. Wheeler and Jay M. Pasachoff for many helpful comments and suggestions and to Dr. H. C. Arp for supplying many of the fascinating photographs used in the text. I also thank Ms. Louise Nelson for her assistance in preparing and typing the manuscript.

William J. Kaufmann, III

1

The Foundations of Gravitational Theory

For thousands of years, human beings have looked up into the night sky and have been filled with awe and wonder. Countless stars stretching from horizon to horizon, the silver moon going through its phases, and planets wandering among the constellations of the zodiac inspired people to take up the study of astronomy. But few people are satisfied with endless observations alone. It is not enough to go out night after night and merely record the positions of the moon and planets. If one sees an eclipse of the moon on February 10, 1971, and another lunar eclipse on January 30, 1972, one is naturally prone to ask when the *next* eclipse will occur. It was apparent to ancient peoples that the motions of the sun, moon, and planets among the fixed stars seemed to follow rhythmic cycles. If one could understand these cycles, one could *predict* where various celestial objects would be in the future.

Only recently have we become fully aware of the ingenuity and depth of insight possessed by ancient peoples in the field of astronomy. The pyramids in Egypt, Stonehenge in England, and the ziggurats of Babylonia are impressive astronomical monuments. In studying these monuments, we realize that for ancient man to have constructed Stonehenge is no less of an achievement than for modern man to have journeyed to the moon.

As far back as 3000 and 4000 years ago, people had devised elaborate systems by which the motion of the sun, moon, and planets could be predicted. The remarkable accuracy of these ancient methods is well known. It has been said that two Chinese astronomers, Hi and Ho, were executed for failing to predict an eclipse in 2159 B.C. Even today with our advanced techniques, the penalties for making an error in one's computations are considerably less severe!

The methods devised by the ancients for calculating the positions of celestial objects and explaining their motions are truly impressive. Eudoxus proposed a system of concentric crystalline spheres with the earth at the center. The sun, moon, and planets were attached to the spheres, which rotated at various rates with respect to each other. Two centuries later, Hipparchus proposed his famous system whereby the planets were attached to epicycles, which in turn traveled about a deferent centered on the earth. This and similar systems were so successful in predicting planetary positions that today one can turn to Ptolemy's *Almagest* and calculate where the moon, for example, will be tonight. Surprisingly, your calculations will agree fairly well with what you observe.

Why, then, have we moderns rejected these various ancient cosmological models? Why did we become dissatisfied with the idea that the earth is at the center of the universe? Questions such as these are of fundamental importance because the entire psychological and philosophical orientation of the Western world is involved. But from a much simpler scientific viewpoint, there are at least two reasons for going from a geocentric cosmology to a heliocentric picture of the solar system.

Perhaps the most obvious motivation for our reorientation resulted from increased astronomical knowledge, beginning with Galileo's use of the telescope. Galileo's discoveries of the phases of Venus, mountains on the moon, sunspots, and the Jovian satellites played an important role in the overthrow of the Aristotelian-Ptolemaic systems. In other words, a Copernican cosmology with the sun at the center of the solar system seems far more physically reasonable.

But with the work of philosophers such as Hume and Kant, it was realized that through physical science we can never hope to gain insight into the *true* nature of reality. The best we can do is prove what is not true. Therefore, there are probably psychological motivations for the changes in astronomy during the Renaissance.

At work deep in the soul of every scientist is a variety of subconscious or even mystical ideas concerning the nature of reality. For example, it is believed that the laws of nature must be beautiful, harmonious, and simple. If someone were to explain a physical law to you, it is believed that you would ideally find such knowledge aesthetically pleasing. Clearly, the work of astronomers who labored under a geocentric hypothesis was not aesthetically appealing. As the accuracy in the determination of planetary positions increased, more spheres or more epicycles had to be added. As the years went by, these systems became more cumbersome, less simple, and frankly less beautiful.

The first major reinfusion of beauty in theoretical astronomy came with the work of Johannes Kepler in the seventeenth century. Kepler's major contributions dealt with a precise description of how the planets move about the sun. Because his work opened the door to modern physical science, it would perhaps be valuable to review the essence of Kepler's discoveries as stated in his so-called Three Laws.

Up until the time of Kepler, all major ideas concerning the motions of the planets used circles. The possibility that other curves might be necessary was never seriously considered. It was Kepler who first proposed that an ellipse (see Figure 1-1) and not a circle must be used in describing the orbit of a planet. This idea is stated in his first law as follows:

Each planet moves about the sun
in an orbit which is an ellipse with
the sun at one focus of the ellipse.

Kepler had at his disposal all the astronomical records of Tycho Brahe, and they constituted some of the finest observations available. Based on these observations, Kepler was then able to

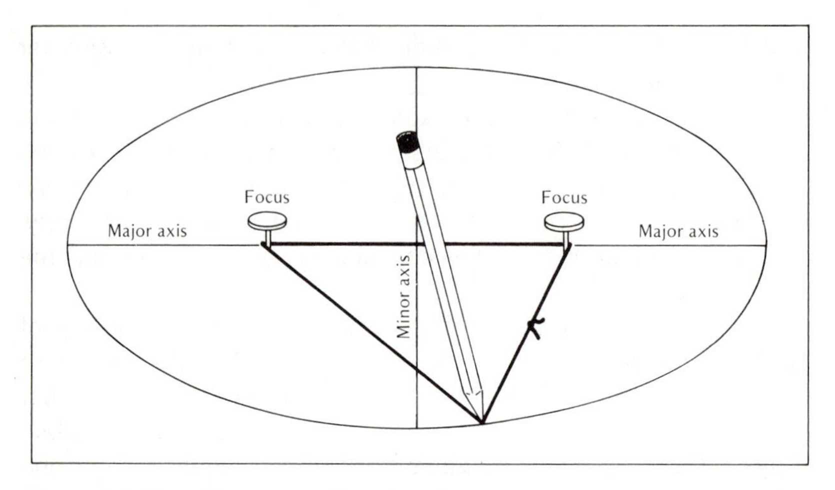

Figure 1-1. The ellipse. An ellipse can be drawn very simply using two thumbtacks, a loop of string, and a pencil, as shown in the diagram. The parts of the ellipse are the two foci (given by the location of the thumbtacks), the major axis, and the minor axis. The major axis is the longest distance across the ellipse and passes through both foci. The minor axis is the shortest distance across the ellipse passing through the center of the ellipse. The semimajor axis refers to one-half of the major axis. Similarly, the semiminor axis is one-half of the minor axis. Note that as the foci get closer and closer, the ellipse looks more and more like a circle.

make a definite statement about how a planet moves along its elliptical orbit. This discovery is formulated in his second law:

> *A straight line joining the sun and a planet sweeps out equal areas in equal time.*

Finally, Kepler's third law relates in a quantitative way the orbital period of a planet to the size of its elliptical orbit as follows:

> *The squares of the periods of planets are in direct proportion to the cubes of the semimajor axes of their orbits.*

The first and second laws are shown schematically in Figure 1-2, while a graph demonstrating the third law is given in Figure 1-3.

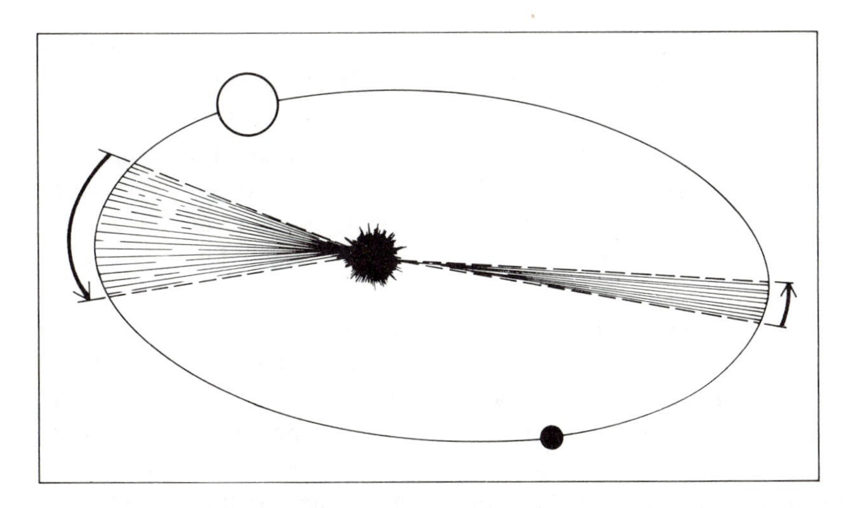

Figure 1-2. Kepler's first and second laws. Each planet moves around the sun in an orbit that is an ellipse with the sun at one focus. In addition, a line joining the sun and a planet sweeps out equal areas in equal times.

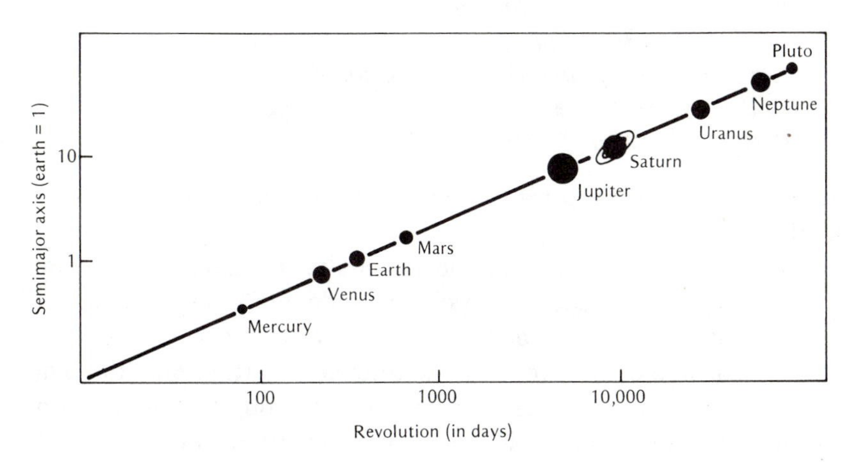

Figure 1-3. Kepler's third law. This graph shows a plot of the period of revolution of a planet about the sun against the length of the semimajor axis of the planet's orbit. The fact that the data from all the planets lie along a straight line tells us that there is a very simple relationship between these two quantities. Kepler's third law predicts the exact slope or orientation of this line.

The planets move through the vast emptiness of space unhampered by many of the extraneous effects that complicate the motion of everyday objects. For example, planets do not encounter friction or air resistance. Clearly, therefore the first major advances in discovering the physical laws of the universe were destined to come from astronomy. In the motions of the planets we see the nature of the physical universe revealed in its simplest and purest form.

It was Isaac Newton who possessed the genius to fathom the meaning of Kepler's Three Laws. To accomplish this, Newton had first to invent new mathematical methods, the differential calculus and the integral calculus, which enabled him to deal with variable quantities such as the distances and speeds of the planets about the sun. By applying these new mathematical tools to Kepler's empirical discoveries. Newton concluded that the planets move about the sun under the influence of a force called *gravity.* More precisely, Newton formulated his theory of gravity as follows:

> *Two bodies exert a gravitational*
> *force on each other which is*
> *proportional to the product of*
> *their masses and inversely proportional*
> *to the square of the*
> *distance between them.*

A graph showing the strength of the gravitational field is given in Figure 1-4.

Every physical object that possesses mass also has a gravitational field. If the mass of the object is doubled, the strength of the field also doubles. In addition, if you are at a certain distance from a massive object, you will experience a certain gravitational force. If you move twice as far away from that object, the strength of the gravitational force will be only one-quarter as great.

Using this law of gravitation and the mathematical methods he developed, Newton then found he was able to derive all of Kepler's laws. In other words, he could explain Kepler's discoveries in terms of an attractive gravitational force that all material bodies (not just planets) must possess. Furthermore, Newton was able to explain and predict a number of phenomena that heretofore had not been understood or known. For example, an ellipse

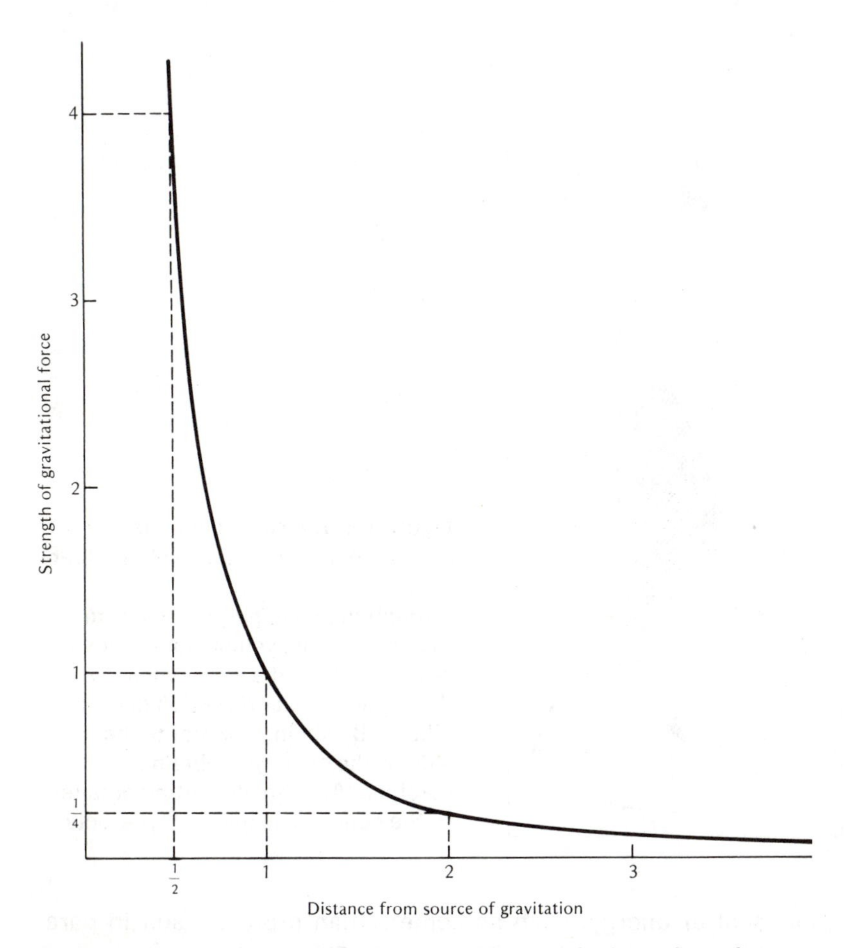

Figure 1-4. Newton's law of gravitation. This graph shows how the gravitational force of a body diminishes with distance from that body. Move twice as far away and the force is only one-quarter as great.

is one member of a family of curves known as *conic sections.* Newton proved that the orbit of an object about the sun could be any one of the conic sections: a circle, an ellipse, a parabola, or a hyperbola. A diagram demonstrating how these curves may be obtained from cutting a circular cone with a plane is shown in Figure 1-5. The precise orbit of a celestial object about the sun would depend in a very specific fashion on how much energy that object possesses. Objects with little energy (compared to the energy contained in the gravitational field of the sun), such as planets, must orbit the sun in circles or ellipses. Objects with a

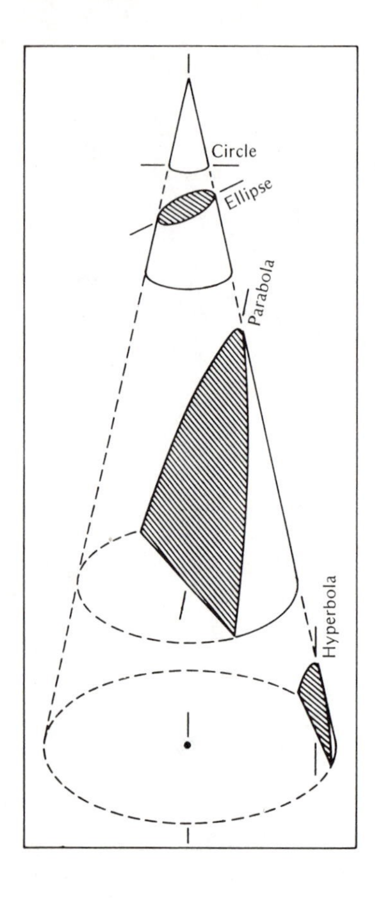

Figure 1-5. The conic sections. By slicing up a cone, you can obtain four very important curves. By cutting through the cone perpendicular to the cone's axis, you obviously get a circle. Cutting at a slight angle to the perpendicular direction gives an ellipse. By slicing parallel to the side of the cone, you obtain a parabola. And, finally cutting parallel to the cone's axis gives a hyperbola.

great deal of energy, such as comets, can orbit the sun in parabolas. These various orbits are shown in Figure 1-6.

Newton's formulation of the first successful theory of gravitation is fundamentally very beautiful. From one simple set of assumptions stated in his universal law of gravitation, he was able to prove the validity of Kepler's laws and explain a number of additional phenomena. In the three hundred years following Newton, the successes of his theory of gravitation mounted steadily. Great mathematicians and physicists such as Gauss, Euler, Lagrange, Laplace, Hamilton, and Jacobi elaborated upon the framework devised by Newton to give us the description of physical reality known as *classical mechanics.* By the turn of the last century, however, a number of troublesome inconsistencies between classical mechanics and electromagnetic theory had become apparent. It was Albert Einstein who resolved these in-

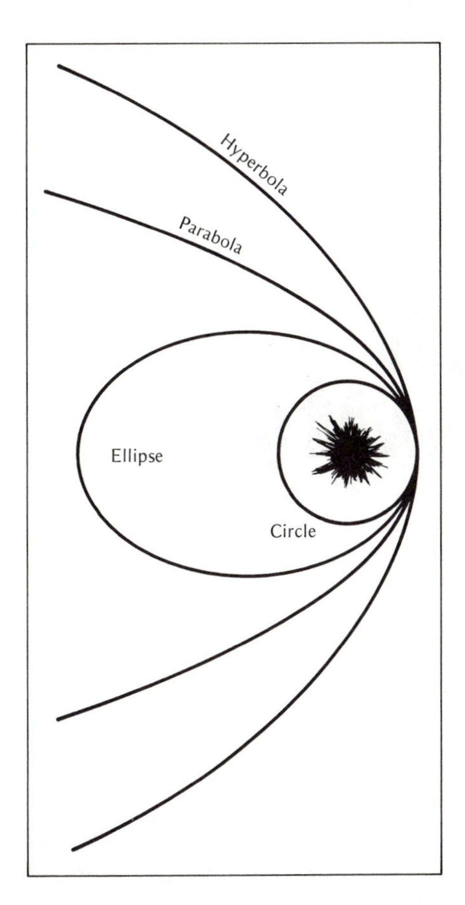

Figure 1-6. Orbits about the sun. The circle is the lowest energy orbit. With more energy, a particle can go into an elliptical orbit and get farther away from the sun. With still more energy, a particle can go into a parabolic orbit and escape from the sun. But at an infinite distance from the sun, the particle would come to rest. If the particle has a very great deal of energy, the orbit will be a hyperbola. Even infinitely far from the sun, the particle would still be moving away from the sun with a considerable speed.

consistencies and then forged ahead to formulate a fundamentally new and different theory of gravitation known as the *general theory of relativity.* As we shall see, Dr. Einstein's theory is even more aesthetically pleasing than Newton's.

When Einstein published has general theory of relativity in 1916, there was a great deal of excitement about observational tests of this new theory. Then, for several decades, interest in general relativity waned. Up until the late 1950s, activity was confined to a small group of physicists and mathematicians who were aware of the implications of Einstein's ideas. Then, with the discovery of quasars, there was a tremendous revival of interest in general relativity; so today there are more theoreticians working in this field than ever before. As a result of recent work, the full potentialities of general relativity are becoming apparent. These potentialities will be explored in detail in later chapters.

The Birth of Relativity Theory

The major advances in physics during the nineteenth century centered about electricity and magnetism. Prior to the work of Oersted and Faraday, it has been thought that electricity and magnetism were two separate, unrelated phenomena about which very little was understood. But as a result of a rash of experiments in the early 1800s, it was well established that these two phenomena are intimately connected. For example, a moving charge gives rise to a magnetic field. And, conversely, a changing magnetic field produces an electric field without the aid of any electric charges.

As a result of these discoveries, scientists were confronted with phenomena that do not fit very well within the framework of traditional Newtonian mechanics. Newton held that only the distance between two objects determined the strength of the forces that they exerted upon each other. But Oersted, for example, discovered than an electric current, which is nothing more than moving electrical charges, will exert a force on the needle of a magnetic compass. This experiment therefore proved that motion is important in determining the nature of an electromagnetic interaction. In addition, while celestial bodies only attract each other, electric forces can be either attractive or repulsive. And, finally, Hertz demonstrated that electromagnetic disturbances move with a finite velocity, whereas prior astronomical observations seemed to indicate that the forces between two bodies depended only on their instantaneous configuration.

In the mid-1800s, the brilliant Scottish physicist James Clerk Maxwell succeeded in describing all these interesting phenomena in one unifying theory. Maxwell accomplished this by emphasizing the concept of a *field.* Whereas Newton held that bodies act directly upon each other across empty space, Maxwell argued that bodies such as charged particles produce stresses in their immediate surroundings. These fields of stresses are therefore the intermediaries between material particles. As a result of the introduction of the idea of a field, Maxwell was able to formulate all *electromagnetic theory* in four simple equations, known as *Maxwell's field equations.*

One of the important predictions of these four equations is that all electromagnetic waves (radio waves, light, x-rays, and so forth) propagate through empty space at a very specific velocity, usually designated by the letter c. This velocity has the well-known value of 186,000 miles per second. Thus, in addition to giving precise mathematical descriptions of the relationship between electric charges and currents and the resulting electric and magnetic fields, Maxwell's field equations for the first time present us with a fundamental velocity.

The significance of a fundamenal velocity in physics cannot be overemphasized. In Newtonian mechanics we are concerned only with accelerations at the basic level. But the existence of a fundamental velocity implied that scientists could experimentally determine a so-called inertial frame of absolute rest. To see what this means, consider the earth moving through space with a velocity v. If we shine a beam of light in the direction of the earth's motion, the light should have the velocity $c + v$. If we point the beam in the opposite direction, the total velocity we measure should come out to be $c - v$. As a result of such experiments, it should be possible to determine precisely how the earth is moving through the universe.

The first people to attempt such an experiment were Albert

A. Michelson and Edward W. Morley, two American physicists. Much to everyone's amazement, it was conclusively shown that regardless of the direction in which you pointed a beam of light, the velocity of the light was *always* the same. In other words, this special number (c = 186,000 miles per second) was even more important than anyone had expected. As a result of this Michelson-Morley experiment, we may in essence conclude that either (1) the earth is not moving (which is nonsense) or (2) there is something very wrong with Newtonian physics.

It was, of course, Albert Einstein who possessed the genius to fathom the significance of these difficulties with the classical Newtonian viewpoint. In doing so, he set physics on a totally new course and completely revolutionized our ideas about matter, space, and time.

To begin with, Einstein made the assumption that *the speed of light is an absolute constant.* In other words, no matter who measures the speed of light, no matter how it is measured and no matter whether or not the observer or the source is moving, one will *always* come up with the value of 186,000 miles per second. This is directly opposed to our common experience with space and time. As a result, our ideas about space and time have to be modified.

The first modification to be introduced is that space and time and not completely separate entities. Indeed, it is advantageous to think of time and space as forming a four-dimensional space-time continuum. That is to say, we treat time as a fourth dimension added to the three usual spacial dimensions (up-down, left-right, forward-back).

To give a clear picture of what is meant by space-time, consider a simple two-dimensional example. Imagine an airplane flight from Los Angeles to New York with brief stops at Denver and Chicago. If we draw a graph on which we plot distance traveled against time, we have a diagram of a two-dimensional space-time. The path of the airplane in this space-time is shown in Figure 2-1.

By a similar construction, we can imagine a three-dimensional or even a four-dimensional space-time. For example, Figure 2-2 shows a three-dimensional space-time depicting a person entering a room and walking first to a lamp and then to a chair.

In drawing diagrams of space-time, physicists find it useful

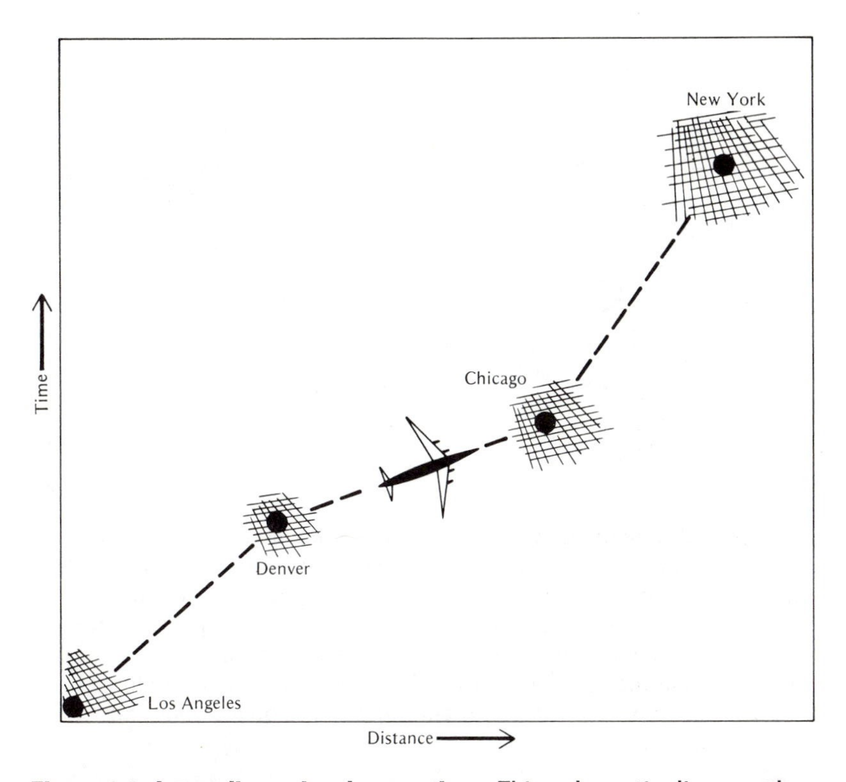

Figure 2-1. A two-dimensional space-time. This schematic diagram shows the flight of an airplane from Los Angeles to New York in space-time. The distance covered is plotted along the horizontal axis, and the time elapsed is plotted along the vertical axis.

to set up their graphs in a very specific fashion. In particular, consider a two-dimensional space-time. Let x be the space dimension and t be the time dimension. Furthermore, set up the graph of x versus t in such a way that light rays travel along 45° lines. This is, of course, a purely arbitrary convention, but it proves to be extremely convenient. Since light rays travel along 45° lines in the two-dimensional space-time, light rays passing through the point $t = 0$ and $x = 0$ define a *light cone,* as shown in Figure 2-3. As a result, space-time is divided up into three regions: past, future, and "elsewhere."

One of the well-known results of Einstein's relativity theory, as we shall see, is that it is impossible to travel faster than the speed of light. Therefore, if in Figure 2-3 the point $x = 0$, $t = 0$ is "here" and "now," it is clearly impossible for observers to know

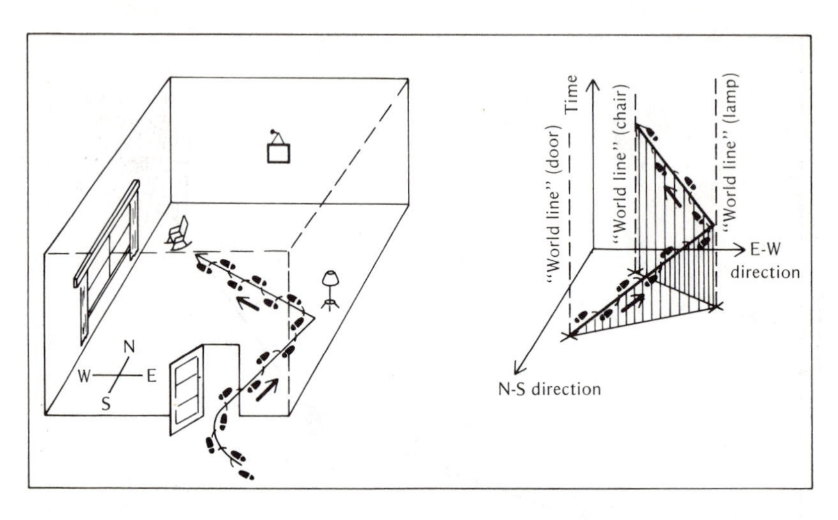

Figure 2-2. A three-dimensional space-time. Suppose you enter a room, walk first to a lamp, and then walk to a chair. Your path in ordinary space is shown on the left. To obtain a diagram of your path in space-time, we must draw a three-dimensional graph. On one axis we plot how far you moved in the north-south direction. On the second axis we plot how far you moved in the east-west direction. On the third axis we plot how much time has elapsed. The resulting diagram giving your path is shown on the right.

anything about the regions labeled "elsewhere." They cannot travel into the "elsewhere" region, because to do so means their spacecraft must go faster than the speed of light, as shown in Figure 2-4. They cannot communicate with anyone in the "elsewhere" region, because to do so means they are sending and receiving signals traveling faster than the speed of light. In addition, to get to "here" and "now" at $x = 0$, $t = 0$, observers must have come from somewhere within the cone labeled "past." And no matter what they try to do, they are destined to be confined to the cone labeled "future" for all time to come.

The true significance of the term *relativity* becomes apparent when we consider two observers who are moving with respect to each other and who try to measure the same thing. For convenience, we shall continue to restrict the discussion to a two-dimensional space-time. In particular, we shall designate distance and time measured by one observer by x and t, while the distance and time intervals measured by a second observer will be called x'

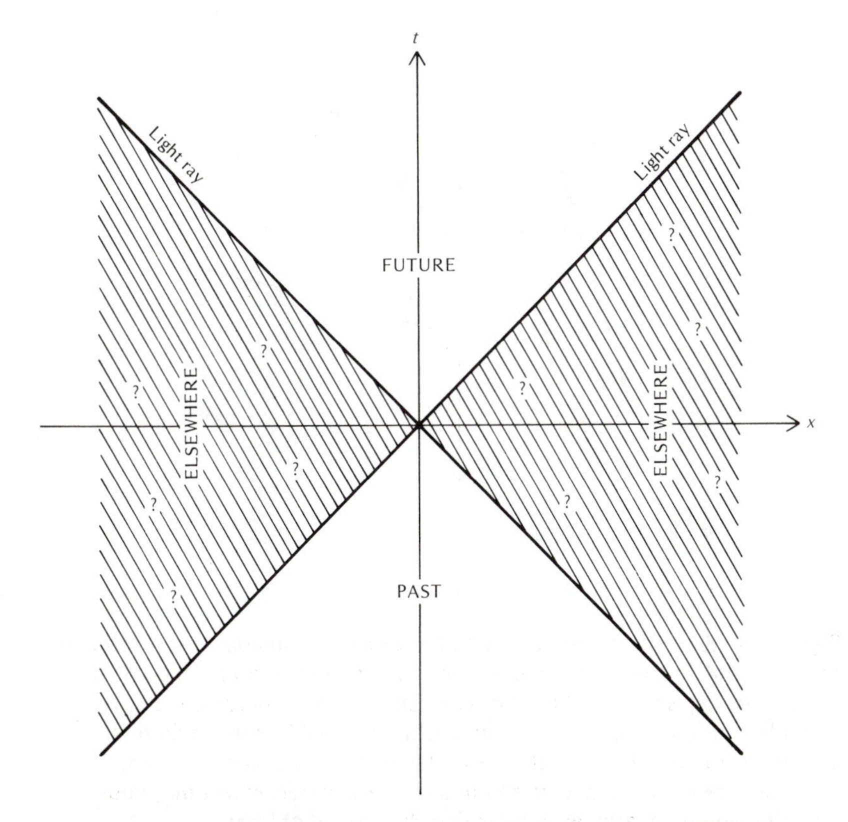

Figure 2-3. The way a scientist draws a two-dimensional space-time. As usual, we plot time (t) vertically and distance (x) horizontally. We set up the graph so that light rays travel along 45° lines. This is easily accomplished by properly scaling the axes: if 1 inch measured along the t axis represents one second, then 1 inch measured along the x axis represents 186,000 miles. If you call the center of the diagram (at $t = 0$, $x = 0$) "here" and "now," then the diagram naturally divides up into three types of regions.

and t'. Furthermore, the relative velocity between the observers will be called v.

In 1905, Dr. Einstein pubished his Special Theory of Relativity, which, to a large degree, deals with what two observers, moving with respect to each other, measure. At the heart of such treatment are some simple mathematical equations known as the *Lorentz transformations,* which relate the measurements of the two observers. These transformations deal with three basic items: distance, time, and mass.

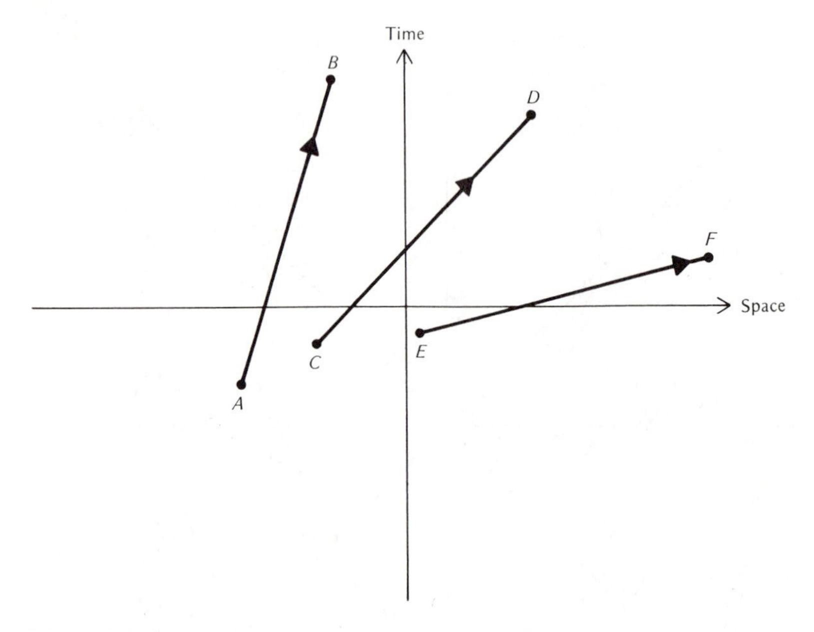

Figure 2-4. Why you cannot get to "elsewhere." Consider three paths in the scientist's drawing of space-time. You can easily go from point *A* to point *B;* a great deal of time passes but not much distance is covered. To go from point *C* to point *D,* you travel along a 45° line; to do this you must travel at the speed of light. In going from point *E* to point *F,* a great deal of distance is covered in a very short period of time; your velocity would have to be greater than the speed of light.

Suppose that we are the observer in the x,t system. Assume that we are at rest and that the second observer (the x',t' system) is moving past us with a velocity v. It then necessarily follows from the constancy of the speed of light that

1. We say that the clocks of the moving observer have slowed down.
2. We say that the rulers of the moving observer have shrunk.
3. We say that the mass of the moving observer has increased.

Of course the moving observer does not notice any of this. He claims that it is he who is at rest; everything is quite normal in his eyes. According to the observer, we are moving, and it is our clocks that have slowed down, our rulers that have shrunk, and our mass that has increased.

At this point we can understand why it is impossible to

travel faster than light. Imagine that an astronaut blasts off from the earth in a rocketship that has an infinite supply of fuel. Month after month, the spacecraft propels the astronaut to higher and higher speeds. However, once the astronaut has achieved a velocity in excess of 75 percent of the speed of light, scientists back on earth begin to notice the slowing of time. According to earth-based scientists, the clocks on board the spacecraft are ticking more slowly than usual. As the astronaut approaches the speed of light, the slowing of time becomes more and more severe. In fact, at the speed of light, the astronaut's clocks appear to stop completely! But the rocket engines may be thought of as clocks. The rate at which they burn fuel may be considered as a means of measuring time. Since time appears to slow down, according to earth-based scientists the rocket engines will consume fuel at a slower and slower rate. Indeed, as seen from earth, the rocket engines will appear to shut off completely at the speed of light. The engines will never consume that final atom of fuel that would propel the spacecraft faster than the speed of light.

Figure 2-5 shows how the mass of a moving object changes with speed. In Figures 2-6 and 2-7 we see how time slows down

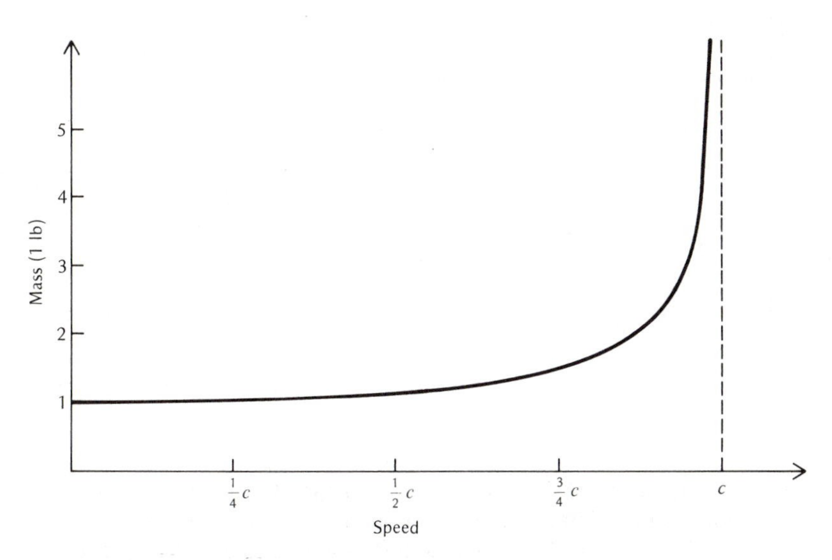

Figure 2-5. Mass increases with increasing speed. This diagram shows how the mass of a 1-pound object appears to increase as the velocity of the object approaches the speed of light.

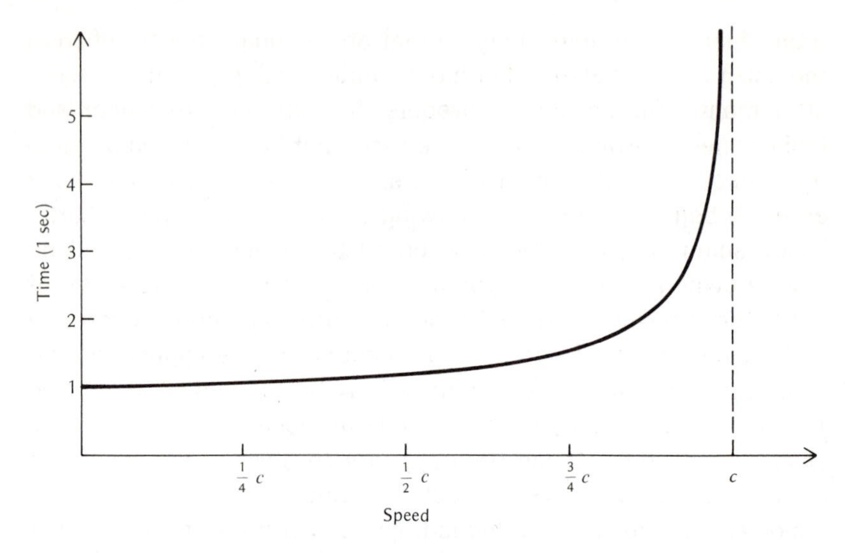

Figure 2-6. Time slows down with increasing speed. This diagram shows how the length or duration of 1 second appears to increase as the clock measuring this 1 second approaches the speed of light.

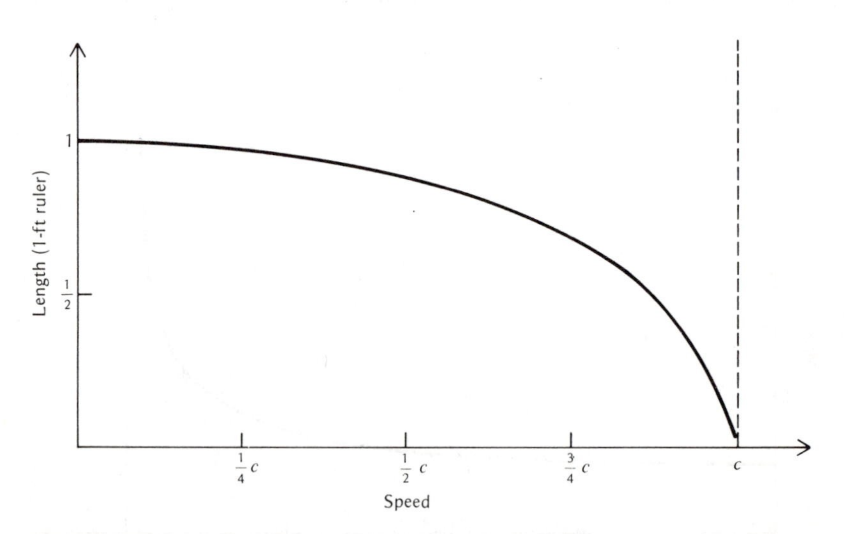

Figure 2-7. Distances shrink with increasing speed. This diagram shows how a 1-foot ruler pointed in the direction of motion appears to shrink as the speed of the ruler approaches the velocity of light. This effect occurs only for distances measured parallel to the direction of motion. A ruler held perpendicular to the direction of motion would remain unchanged.

and rulers shrink. It should be noticed that these effects become important only for extremely great speeds. In fact, one of the only places where these effects become important is in nuclear accelerators, where protons or electrons travel very near the speed of light. In experiments by nuclear physicists, these effects have been confirmed to a very high degree of accuracy.

Finally, it is of interest to notice how the Lorentz transformations affect space-time diagrams. If we say that we are at rest in our x,t system and that the x',t' system is moving, then, in plotting the x',t' system as in Figure 2-3, we find that the x',t' axes are tilted. This is shown in Figure 2-8. This tilting is symmetrical about the light ray line only if the light ray line is at 45°.

At this point it is possible to appreciate why there is no such

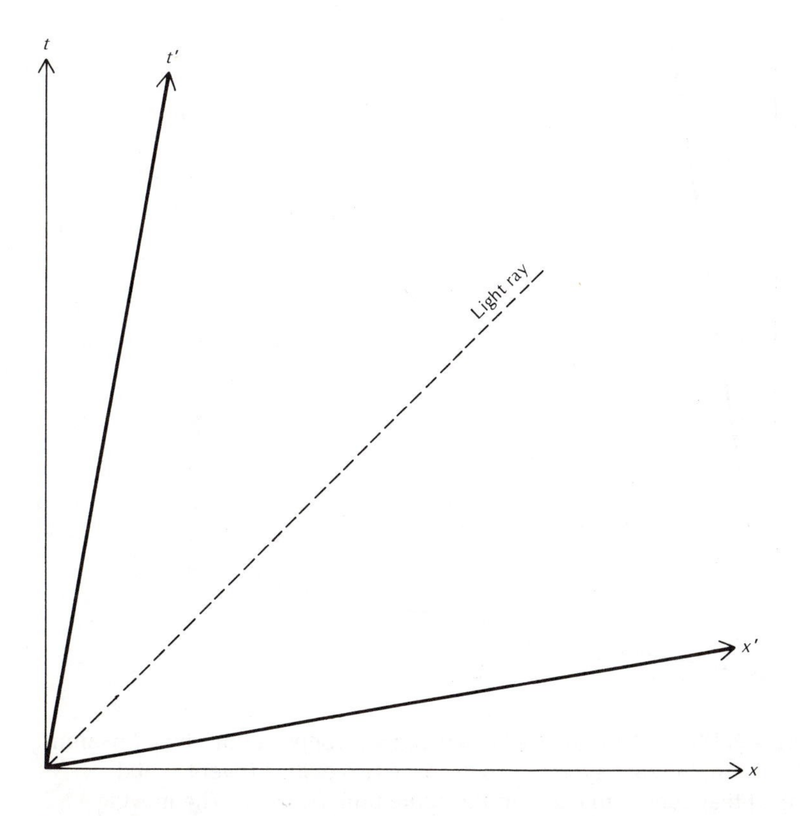

Figure 2-8. How space-time changes for a moving observer. Suppose that the x,t system is at rest. As usual, the t axis is perpendicular to the x axis. If the x',t' system is moving, then its axes are tilted toward the 45° light ray line.

thing as "simultaneity" between two observers moving with respect to each other. Suppose we in an x,t system observe two events that appear to occur at the same time but that are separated by a large distance. Then the time of occurrence of event A equals the time of occurrence of event B, or $t_A = t_B$ in our system. But a moving observer (the x',t' system) will say that these two events did not occur at the same time. The reason for this is most clearly shown in Figure 2-9. Due to the fact that the axes of the x',t' system are titled with respect to the x,t axes, the more distant event seems to occur first according to the moving observer.

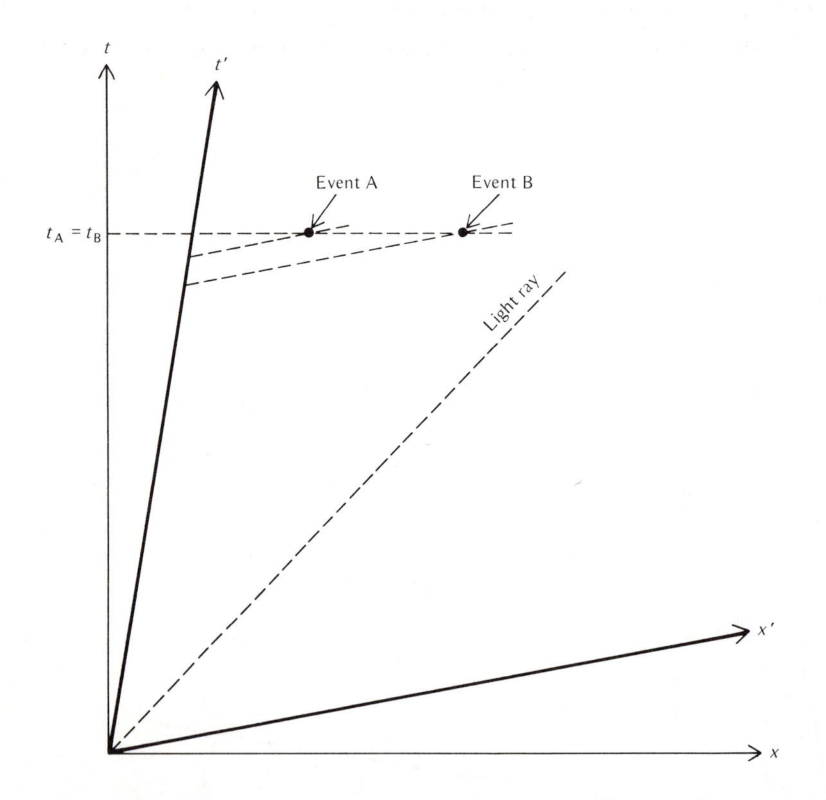

Figure 2-9. Why "simultaneity" is nonsense. Suppose that the x,t system is at rest. In this rest system, we observe two events (event A and event B) that appear to occur at the same time ($t_A = t_B$). The moving observer in the x',t' system will say that the two events did not occur at the same time. Because the axes of the moving system are titled as shown in the diagram, the observer will say that event B occurred before event A.

In these few pages, we have briefly touched upon the highlights of the special theory of relativity, a theory that revolutionized our thinking about space and time. As we shall see, the general theory of relativity, formulated by Dr. Einstein a decade after his special theory, was to have an equally dramatic effect on our thinking about the universe as a whole.

The Foundations of General Relativity

In 1905, Albert Einstein proposed a series of new ideas concerning the nature of space, time, and matter. This set of ideas, known as the *special theory of relativity,* tells what happens when traveling very near the speed of light. During the late 1800s, it became apparent that there were serious difficulties with traditional concepts of space and time. In particular, the work of James Clerk Maxwell provided a much improved understanding of the nature and properties of light. As a result of this understanding, concepts of space and time were clearly inadequate and in need of revision. Dr. Einstein accomplished this revision by assuming that the speed of light is an absolute constant. In addition, he found it especially advantageous to include time as a fourth dimension, along with the three traditional spatial dimensions. All the usual predictions of special relativity then follow as a consequence of the constancy of the speed of light and a four-dimensional space-time continuum.

We can, however, aspire to much more than is presented in the special theory of relativity. We observe, for example, that gravitation is one of the most important forces in nature. It is through gravitation that the planets remain in orbit about the sun. Gravity controls the motion of the sun and the stars in our galaxy. Gravitation dominates the interaction between galaxies and perhaps the evolution of the universe as a whole. It certainly would be extremely enlightening if gravitation could be incorporated into the four-dimensional concept of space-time.

The astronomer is strongly motivated to think of the universe in terms of a four-dimensional space-time. In a very real sense, as we look out into space, we are looking backward in time. If we see a star that is twenty light years away, then the light that is striking our eyes left the star twenty years ago. In other words, we are seeing the way the star looked twenty years past. It is obvious, therefore, that whenever we think about astronomical objects, we find that our ideas about space and time are intimately related. Especially in view of the fact that gravitation is the dominant interaction between astronomical objects, there is a very real motivation for expanding or generalizing the special theory of relativity to include gravity.

But how can this generalization be accomplished? Perhaps the most fruitful approach entails the assumption that the nature of space-time is altered in the presence of a gravitational field. In other words, in gravity-free conditions, particles and beams of light travel in straight lines. In a gravitational field, particles and light rays still try to travel in the most efficient fashion, but because space-time is altered, the resulting trajectory is no longer a straight line. The problem then becomes one of describing in what way space-time is altered and by how much.

In the special theory of relativity, which does not include gravitation, space-time is "flat." Perhaps in the presence of a gravitational field, space-time becomes "curved." A weak gravitational field would cause space-time to be slightly warped, and the trajectories of particles and light rays would deviate slightly from classical straight lines. An intense gravitational field would cause space-time to be severely warped, resulting in major changes in the paths followed by particles and light rays.

To appreciate the significance of this type of approach to gravity, we must obviously have a very clear idea about what is meant by a "flat" or curved" space. It is extremely useful, for purposes of illustration, to restrict the discussion to two-dimensional surfaces. Then, if we have set up our ideas in the most convenient and logical fashion, we should be able to expand the concepts to three-dimensional, four dimensional, or even infinite-dimensional spaces.

We all have a very fine intuitive concept of whether or not a two-dimensional surface is "curved" or "flat." The floor is flat. A table top is flat. But the surface of a basketball or a football is curved. In order to be able to formulate these intuitive observations in a useful fashion, we shall define something called a *vector.* A vector is anything that has both size and direction. Velocity is a good example of a vector; on a freeway you might be traveling with a velocity of 55 miles per hour toward the east. We may symbolize a vector by an arrow. The length of the arrow describes the size or magnitude of the vector, while the direction in which the arrow is pointed gives the orientation of the vector. For example, we might visualize a weather map covered with little arrows describing wind velocity on a particular day. The size of the arrow would reflect the wind speed, and the orientation of the arrow would show the direction in which the wind was blowing. At a glance it is possible to get a clear picture of how the air was circulating on that day.

Now imagine a vector (symbolized by an arrow) on a two-dimensional surface. We then proceed to move this vector around on this surface from place to place without changing its size or orientation. This process, known as *parallel transport,* is accomplished as follows. Suppose you want to move the vector from point *A* to point *B* along a particular path. At each stage along the way you stop and examine the vector, and from step to step along the path you make sure that the size and direction of the vector are unchanged.

We now notice an extremely interesting fact, as shown in Figure 3-1. In going from point *A* to point *B* on a flat surface, you always end up with the same result at the end of the journey, regardless of which path you took to get from *A* to *B.* But on a curved surface, when you get to point *B* you will have a different vector, depending on which path was elected. In other words, in flat space the end result of parallel transport of a vector is independent of the path taken. But in curved space, the end result of parallel transport of a vector depends very critically on which path is taken. This criterion of establishing whether or not space

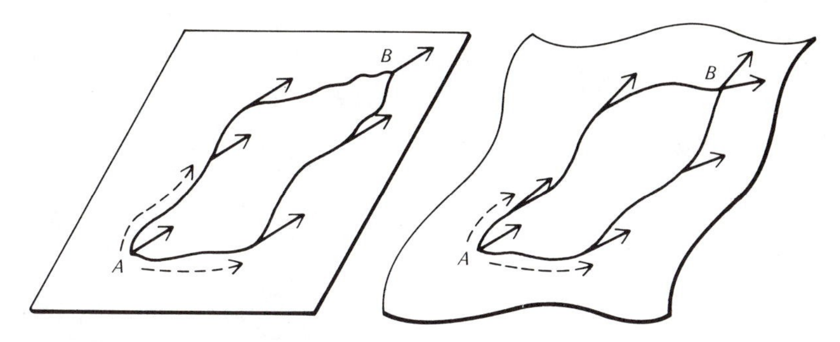

Figure 3-1. Parallel transport of vectors on flat and curved surfaces. In the parallel transport of vectors (symbolized by arrows), when you move along a path, you do your very best from step to step to ensure that the length and orientation of the vector remain unchanged. In going from point *A* to point *B* on a flat surface, you always obtain the same result at the end of the trip. This is independent of the path taken. On a curved surface, however, you will obtain a different final result depending on which path is chosen.

is curved can be extended to spaces of any number of dimensions.

A final, very clear example of this phenomenon is given by someone moving a vector about on the earth. Suppose you start on the equator with a vector of such-and-such length pointed directly east. You then move along the equator through 90° of longitude. From there you go up to the north pole. And from the north pole you return to your starting point. Throughout the entire lengthy trip you are extremely careful that—from step to step—your vector never changes the direction in which it is pointing. As shown in Figure 3-2, by the time you get back to your starting point on the equator, your vector is pointing toward the north, not the east.

The use of parallel transport of vectors constitutes a very precise means of describing a curved space. In the nineteenth century, mathematicians such as Georg F. B. Riemann, Elvin Bruno Christoffel, and Gregorio Ricci developed the complete theory of curved space of any dimension. Up until the early 1900s, this branch of mathematics was nothing more than a clever toy. But it was Albert Einstein who first clearly realized how extremely useful this mathematics could be. These mathematical tools were ripe for the picking; he had only to give the correct physical interpretation to abstract algebraic symbols.

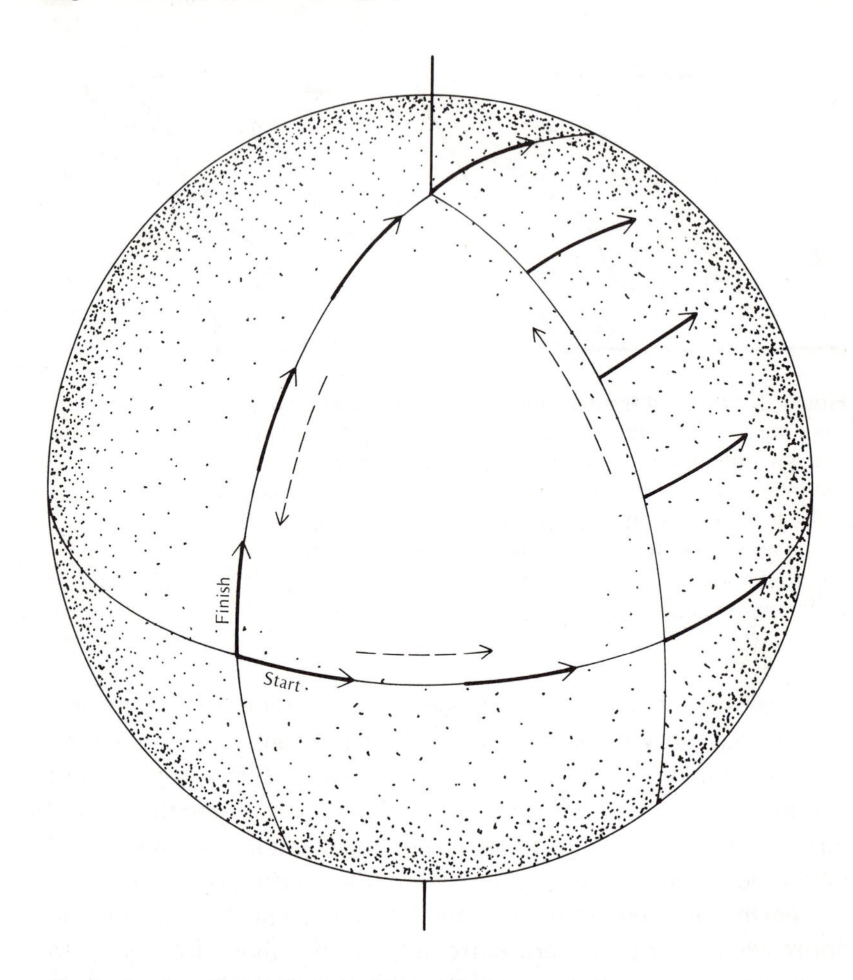

Figure 3-2. Parallel transport of a vector on a sphere. Suppose you start on the equator of the earth with a vector pointed directly east. Parallel transport this vector through 90° of longitude. Then go up to the north pole. Now return to your starting place. At every step along the way, you are very careful that the vector is pointed in the same direction as in the previous step. In spite of this, at the end of your trip the arrow is pointing north, not east!

When Newton first developed classical mechanics, he had to make use of a new type of mathematics called *the calculus.* When Maxwell developed electromagnetic theory, he had to use a new branch of mathematics called *vector analysis.* So also when Einstein began formulating the general theory of relativity, he utilized the recently invented *tensor analysis.* Tensor analysis is the branch of mathematics dealing with mathematical quantities called tensors. By the early 1900s, Riemann, Gauss, Ricci, and Christoffel had discovered a number of tensors that were extremely useful in describing the nature of a curved space. In particular, the so-called *Ricci tensor* tells us how vectors turn as you parallel transport in a curved space. This is exactly what Einstein had been looking for. From the Ricci tensor, Einstein then constructed the *field equations.* These all-important equations tell us all we need to know about the properties of space-time in a gravitation field.

But clearly this is not enough. The Einstein field equations tell us how space-time is warped by a gravitational field, but they do not tell us how objects move in that curved space. To deal with this situation, we assume that particles and light rays travel in the most efficient fashion through space-time. To be more precise, if an object is to travel from point *A* to point *B,* the trajectory that it follows will be the shortest distance connecting those two points in space-time. If space-time is flat, clearly the trajectory will be a straight line. If space-time is curved, the resulting, most efficient, trajectory is called a *geodesic.* The equations one must solve to learn the path followed by an object or light ray are called the *geodesic equations.* Many of the problems in general relativity, therefore, consist of simultaneously solving the field equations, along with the geodesic equations. In the following chapters many of the fascinating results of such calculations will be examined.

4

Experimental Tests of Relativity

By the turn of the last century, it had become increasingly obvious that there was something wrong with the motions of the planet Mercury. From Newtonian mechanics, astronomers should have been able to calculate very precisely the orbits of all the planets. But even after all known corrections due to disturbances by the outer planets had been applied, the calculations failed to predict the movement of Mercury with anticipated accuracy. In particular, Mercury's elliptical orbit semed to be moving around, or *precessing,* very slowly. In the absence of the influence of the outer planets, Mercury's orbit, according to classical mechanics, should have been a perfect ellipse. But observations revealed that while Mercury revolved about its orbit, the orbit was slowly rotating. As a result, the planet Mercury was tracing out a rosette figure rather than a true ellipse, as shown in Figure 4-1.

This effect is extremely small. The perihelion of Mercury's orbit (i.e., the point in the orbit nearest the sun) moves through an unexplained 43 seconds of arc each century. Yet this so-called precession of Mercury's perihelion was deeply disturbing to astronomers in the early 1900s. They thought that they should have been able to account for it.

In 1916, Albert Einstein presented a new and revolutionary theory of gravitation known as the *general theory of relativity.* The central idea of this theory is that a gravitational field is manifested by a curvature or warping of four-dimensional space-time. The so-called *field equations* tell how severely space-time is curved, and the *geodesic equations* tell how objects move in that warped space. When Einstein turned his attention to the motion of planets around the sun or a massive star, he found that ideally the orbits should, in fact, be precessing ellipses. Near the sun or a star, where the gravitational field is most intense, the rate of precession should be the greatest. Far from the source of gravity, where the field is weak, any deviations from Newtonian mechanics should be virtually unnoticeable. Indeed, according to Einstein's calculations, the rate of precession of Mercury's perihelion sould be 43 seconds of arc per century. For the first time we had an explanation for the mysterious motion of Mercury, and the explanation required us to change all our ideas about gravitation.

It is perhaps difficult to convey the full extent of the excitement that spread through the scientific community in 1916. Newton's classical theory of gravitation had stood as an unshakable pillar of physical science for centuries. And here suddenly was a totally new and different theory that seemed to work better!

The immediate task of scientists at this point was to test this

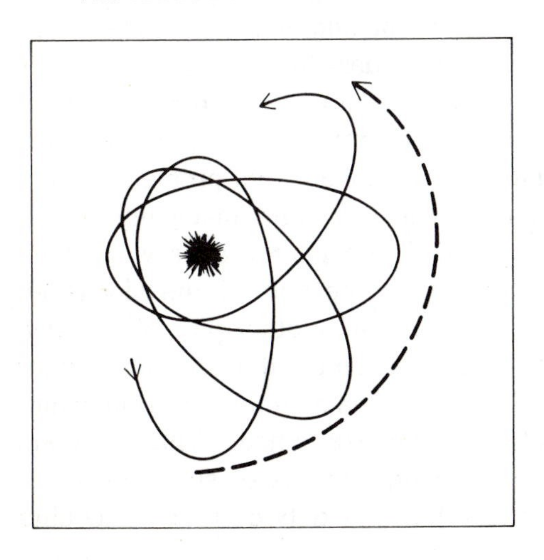

Figure 4-1. The precession of Mercury's perihelion. After subtracting the extraneous effects due to the outer planets, according to classical theory, Mercury's orbit about the sun should be a perfect ellipse. Instead, the ellipse itself moves slowly, or precesses. As a result, Mercury traces out a rosette figure. (In the diagram, this effect is highly exaggerated.)

new theory as much as possible. Unfortunately, for weak gravitational fields Einstein's theory predicts almost exactly the same results as the old Newtonian theory. The only place where gravitational fields are not very weak is near the surface of the sun. Setting up experiments to check general relativity is therefore not an easy job.

The precession of Mercury's perihelion, which had been known since the time of Leverrier in the 1800s, provided the first "test" of general relativity. In essence, near the sun, the curvature of space-time is so large that planets are no longer able to travel in simple ellipses. Rather, while the planet attempts to revolve about the sun or a star in the elliptical orbit predicted by Newton, the ellipse precesses. The stronger the gravitational field, the faster the rate of precession. Because the rate of precession in the case of Mercury is so small, we see that, from the viewpoint of general relativity, the gravitational field of the sun is comparatively weak.

A second test of general relativity is provided by the fact that light rays pass very near the sun. As beams of light from distant stars pass thruogh the curved space-time near the sun's surface, the light rays should be bent or deflected slightly, as shown in Figure 4-2. The idea that light rays could be affected by a gravitational field had, up until now, never been seriously considered. But Einstein's calculations revealed that a beam of light just grazing the sun's surface should be deflected through an angle of 1.75 seconds of arc.

Obviously, the difficulty with making such an observation stems from the fact that the sun is so dazzlingly bright that, under normal conditions, it is impossible to view stars in the sun's vicinity to see if they have moved. The total solar eclipse of 1919 provided a unique opportunity to circumvent this problem. Photographs of the stars surrounding the sun were taken during the few precious minutes of totality. These photographs were then compared with photographs of the same part of the sky taken six months earlier, when the sun was in another part of the sky. Although these observations were extremely difficult and fraught with inaccuracies, Einstein's theory again emerged triumphant. During almost every total solar eclipse since 1919, someone repeats this experiment with varying degrees of success.

We therefore have a theory that not only correctly accounts

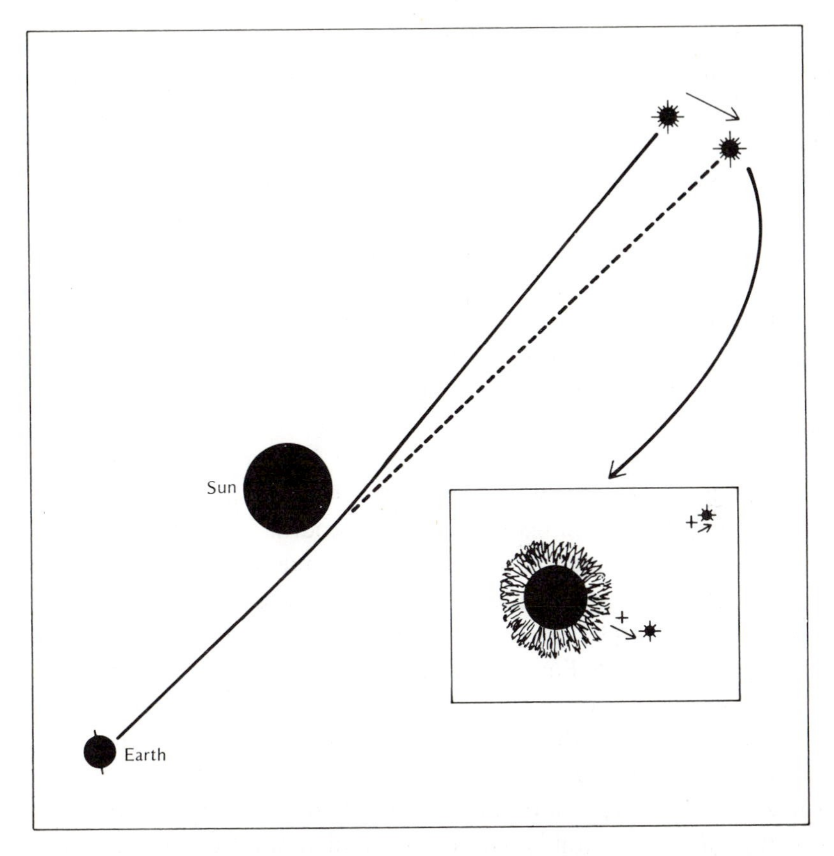

Figure 4-2. The deflection of light by the sun. A light ray passing near the surface of the sun is deflected slightly. This effect was first observed during a total solar eclipse.

for phenomena that previously had been unexplainable but also predicts new phenomena that heretofore had been unknown.

A second unexpected prediction from the general theory of relativity has to do with the behavior of clocks in a gravitational field. A clock in an intense gravitational field goes slower than a clock far from a massive body. Indeed, time slows down in a gravitational field. This effect, known as the *gravitational redshift,* was tested only very recently. In the early 1960s, three physicists at Harvard University, Drs. Pound, Rebka, and Snyder, made use of the so-called Mössbauer effect to test the slowing of clocks in a gravitational field. In a very real sense, atoms and radioactive nuclei are clocks. The Mössbauer effect permits scientists to examine very accurately the frequency of gamma rays emitted by certain types of radioactive nuclei. The experiment was performed in a tall building. According to general relativity, clocks on the ground floor or basement of a building should tick more slowly than clocks on the roof; the ground floor is nearer to the center of the earth and therefore in a more intense gravitational field. By comparing the frequency of gamma rays used in the Mössbauer effect in the basement and several stories higher up, the Harvard physicists succeeded in detecting a slowing of time, as shown schematically in Figure 4-3. Their results agree with the predictions of Einstein's theory, with an accuracy of better than 1 percent.

These three effects (the precession of Mercury's perihelion, the deflection of light by the sun, and the gravitational redshift) constitute the *three classical tests* of general relativity. In a certain sense, the expansion of the universe may be considered as a fourth test, but the three classical tests are really the only experimental justification we have for Einstein's theory. This is extremely unfortunate. For more than half a century, physicists have produced a deluge of books and papers on a theory for which, in the strictest sense, there is little experimental justification.

The situation became extremely critical in the recent past. A number of physicists proposed new theories of gravitation that are similar to Einstein's work but fundamentally different. For example, Drs. Brans and Dicke proposed that a weak *scalar field*, in addition to the usual Einsteinian *tensor field,* might be important in gravitation. The predictions of the Brans-Dicke scalar-tensor theory are almost identical with those of traditional general relativity. Although the Brans-Dicke theory is the most famous of the non-Einsteinian general relativity theories, there are several additional formulations that are fundamentally different from Einstein's work but predict only slightly different results for the three classical tests. The problem then becomes one of obtaining very accurate tests of general relativity. The only test that can be done in the laboratory, namely the gravitational redshift experiment using the Mössbauer effect, is of no help. All non-Einstenian theories predict, for one reason or another, the same results for this experiment.

The deflection of light past the sun, measured during a total

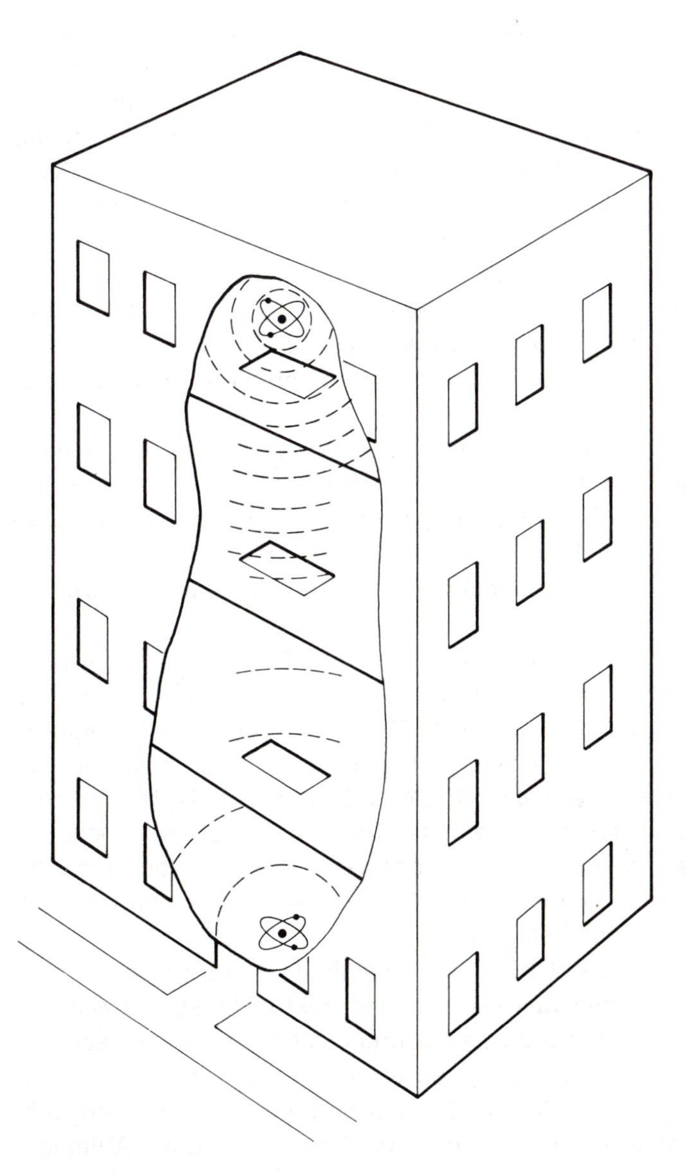

Figure 4-3. Gravitational redshift. According to general relativity, clocks slow down in a gravitational field. Therefore, a clock on the ground floor of a building wil run slow compared to a clock in the attic, which is farther away from the center of the earth. This effect was first measured using gamma rays from radioactive nuclei in a house near Boston.

solar eclipse, is a very difficult observation. Eclipses often occur over oceans or jungles, which are inaccessible to scientific equipment. Frequently there are no bright stars near the sun at the time of the eclipse. And the brief duration of the eclipse means that if something goes wrong, you cannot do it over again. As a result, the measurements of the deflection angle of a beam of light just grazing the sun's surface are somewhat uncertain. The possible errors due to inaccuracies in the measurement are at least 8 percent. For example, during the eclipse of 1952, this angle was measured to be 1.70 seconds of arc, with an error of plus or minus 0.10 second of arc. Einstein predicted 1.75 seconds of arc.

Clearly, the agreement with Einstein's prediction is dramatic. But many of these newer theories predict results around 1.60 or 1.70 seconds of arc, for example. Obviously, they cannot be ruled out. Unfortunately, there is little hope of significantly improving the classical eclipse observations. Scientists must turn to something else.

One possibility involves the fact that the sun emits very little microwave radiation. But other sources near the ecliptic, such as the quasars 3C 273 and 3C 279, are very bright in microwaves. Every October 8, the sun eclipses 3C 279. Therefore, it should be possible to measure the deflection of light (i.e., microwaves) without having to wait for the usual total solar eclipse. The only immediate problem is improving the resolution of microwave telescopes. The experimental error would then be reduced to plus or minus 1 percent or less.

For many years it had been thought that the precession of Mercury's perihelion was the strongest test of Einstein's general relativity. His theory predicted a precession rate of 43 seconds of arc per century, which agrees to within about 1 percent with actual observations. Most of the newer theories predict considerably less, such as 38 or 40 seconds of arc per century. Although this confirmation of Einstein's ideas seemed iron clad, all of the work with orbit theory in the solar system assumed that the sun was perfectly spherical. In the late 1960s, Dr. Dicke at Princeton University performed a series of elaborate and ingenious observations that suggest that the sun is in fact an oblate spheroid. It is fatter at the equator than at the poles. If the sun is oblate, ordinary Newtonian mechanics tells us that an equatorial bulge will produce a small amount of precession having nothing at all to do with relativity. In other words, of the observed 43 seconds of arc per century, 4 or 5 seconds of arc might be due to ordinary classical mechanics. Therefore, relativistic calculations must account for only 38 or 40 seconds of arc per century.

The improvements with regard to the observations of Mercury will be twofold. First, the oblateness of the sun must be examined in detail. If the surface of the sun appears oblate, does this mean that the entire gravitational field of the sun is that of an oblate spheroid? Second, scientists must have better knowledge of the location of the planet Mercury itself. The best way to accomplish this is to place an artificial satellite in orbit about Mercury. This will probably be done during the normal course of the space program in the coming years.

The space program itself offers many fine opportunities to test general relativity. Because time slows down in a gravitational field, the radio transmissions from spacecrafts passing behind the sun are delayed. The observed amount of delay agrees with the Einstein theory to within 3 percent. One of the dfficulties is that the hot gases near the sun, the *solar corona,* also slow down radio waves. Using transmissions at only one frequency, it is not easy to differentiate the effects due to the corona from the effects due to gravitation. If a future spacecraft destined to fly behind the sun is equipped with transmitters capable of transmitting at several frequencies, this difficulty will vanish. Gravitation should affect all light of all frequencies in exactly the same fashion. Any difference between the transmissions at separate frequencies is, therefore, due entirely to the gases in the solar corona. We may then distinguish clearly between the effects of the corona and those of gravitation.

Another prediction of relativity that could be tested with space technology has to do with the behavior of gyroscopes. As a body such as the earth rotates, to a small degree space-time should be dragged along around the rotating earth. This so-called dragging of inertial frames would cause a gyroscope in an orbiting satellite to precess very slowly. The rate of precession could easily be used as a test of general relativity.

By way of summary, the tests of relativity theory must be

improved to the extent that the experimental errors are reduced to less than 1 percent. Only then will we be able to know with confidence whether or not Einstein's ideas are really on firm ground. It should, however, be pointed out that, mathematically, Einstein's theory is extremely simple and beautiful. The competing theories are not. If beauty and simplicity are in some way a measure of validity, we may continue with confidence in assuming Einstein was right.

The Meaning of the Redshift

Everything known about the universe, with the exception of what is known about the earth and the moon, comes from the study of light. The only way we know anything about the universe beyond the earth's domain is through the examination of electromagnetic radiation, whether it be visible light, radio waves, or x-rays emitted by the stars and galaxies. A clear understanding of the nature of the universe, therefore, presupposes an intimate knowledge of the properties and behavior of light.

It is well known that if a beam of white light is passed through a prism, the white light is broken up into the colors of the rainbow, as shown schematically in Figure 5-1. By examining this rainbow, or spectrum, a great deal can be learned about the nature of the original source of light. For example, close inspection of the spectrum may reveal a number of lines interspersed among the colors. These spectral lines can be used to identify the particular atoms in the source of light. One chemical element gives one set of spectral lines while atoms from any other element give an entirely different set of spectral lines. In a very real sense,

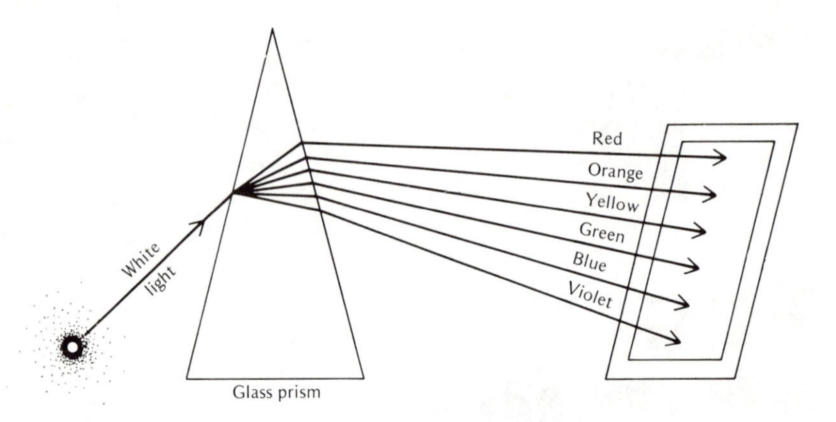

Figure 5-1. The prism. When a beam of white light is passed through a prism, the light is broken up into the colors of the rainbow. If the light from a star is passed through a prism, in addition to the colors of the rainbow, we see many dark lines among the colors. These dark lines (spectral lines) are formed by the atoms in the star's atmosphere.

the spectral lines are the fingerprints or signatures of various atoms or chemicals in a source of light. The chemical composition of stars and galaxies can be discovered simply by identifying the lines in the spectra of these objects. No two different elements can give rise to the same pattern of spectral lines. These patterns are determined in simple laboratory experiments; and the identification of the lines is, therefore, in principle, unambiguous.

By examining spectral lines, astronomers can learn a great more than just the chemical composition of astronomical objects. Various conditions at a source of light cause the spectral lines to be distorted in specific ways. By detailed examination of these distortions (broadening or splitting of spectral lines), astronomers can, for example, determine how fast a star is rotating or how hot the surface of the star is. They can discover the pressure in the star's atmosphere or the strength of the star's magnetic field. The wealth of information about the universe contained in starlight is truly immense. Scientists are limited only by the quality of their instruments and the ingenuity of their observations.

Back in the nineteenth century it was realized that the motion of a source of light has a significance effect on the positioning of spectral lines. It is our common experience to have listened while an ambulance or fire engine passed us on a street. While the ambulance is approaching us, the pitch of the wail from the siren seems high. When it moves away from us, the pitch from this same siren seems noticeably lower. In other words, the frequency, or pitch, of a tone emitted by a source approaching us appears high, while the pitch of the same tone appears lower on on a receding source. This is known as the *Doppler effect.*

This same effect also occurs with light. If a source of light is approaching us, the wavelength of the spectral lines in its spectrum are shortened, as shown in Figure 5-2. The entire spectrum is shifted toward the blue. If a source of light is moving away from us, the wavelength of the spectral lines increases, and the spectrum is shifted toward the red. The greater the speed, the greater the shift. By measuring how much the spectral

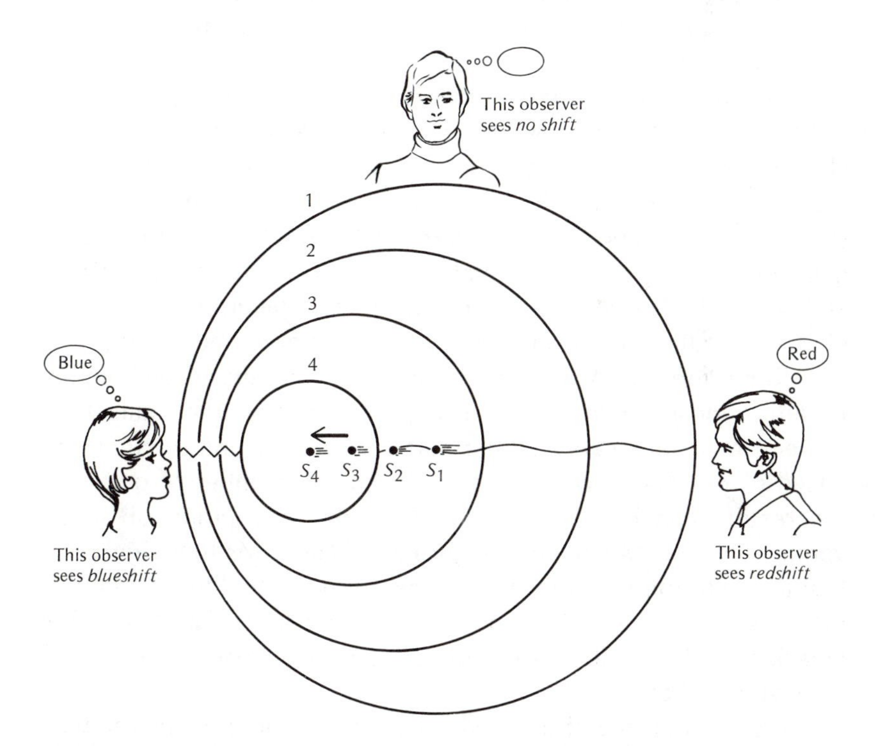

Figure 5-2. The Doppler effect. If a source of light is moving toward you, the wavelength of the light will appear to be compressed. As a result, the light will appear shifted toward the blue end of the spectrum. If the source of light is moving away from you, the waves will appear to be spread out. This will result in the light being shifted toward the red end of the spectrum.

lines are displaced, we can easily determine how fast a source of light is traveling.

This Doppler effect has important astronomical applications. By examining the displacements of spectral lines in stellar spectra, astronomers can determine whether a star is moving toward us and at what speed. We can measure how fast a spherical shell of gas is expanding from a star that has undergone a novalike explosion. We can go a long way in determining the orbit of a binary star because from the Doppler shift we know the velocity of the stars as they move about their common center of mass. From the shifting of lines in the spectra of clouds of hydrogen gas, we can tell how they are moving, which, along with orbital theory, gives us a picture of the structure of our galaxy as a whole.

Perhaps one of the most dramatic discoveries to come about through the use of the Doppler effect deals with the nature of the universe as a whole. Back in the 1930s, the American astronomer Edwin Hubble took spectra of many different galaxies. With the exception of the very nearest galaxies, all the spectra showed lines shifted toward the red. In particular, he found that galaxies that are thought to be very distant because of their small dim appearance had very large redshifts. On the other hand, galaxies that appear to be relatively nearby had very small redshifts. In other words, nearby galaxies are moving away from us slowly, and distant galaxies are receding from us more rapidly.

To be very precise, astronomers use a variety of techniques to determine the distance to a particular galaxy. They also take a spectrum of this galaxy, determine its redshift, and calculate its speed of recession. When this is done for a number of galaxies, the results can be displayed in the form of a graph of speed versus distance as shown in Figure 5-3. Except for those galaxies that are extremely nearby (e.g., those in our own "local group"), the data from these observations lie along a straight line. This means that if you double the distance to a galaxy, the speed of recession will also double.

The significance of this basic discovery cannot be overemphasized. Indeed, this direct proportion relationship between the distance and recessional speed of far-away galaxies, called the *Hubble law,* is one of the first basic clues to the overall behavior of the universe. To see the meaning of this discovery, we shall consider Dr. Abell's analogy, which involves a grandmother bak-

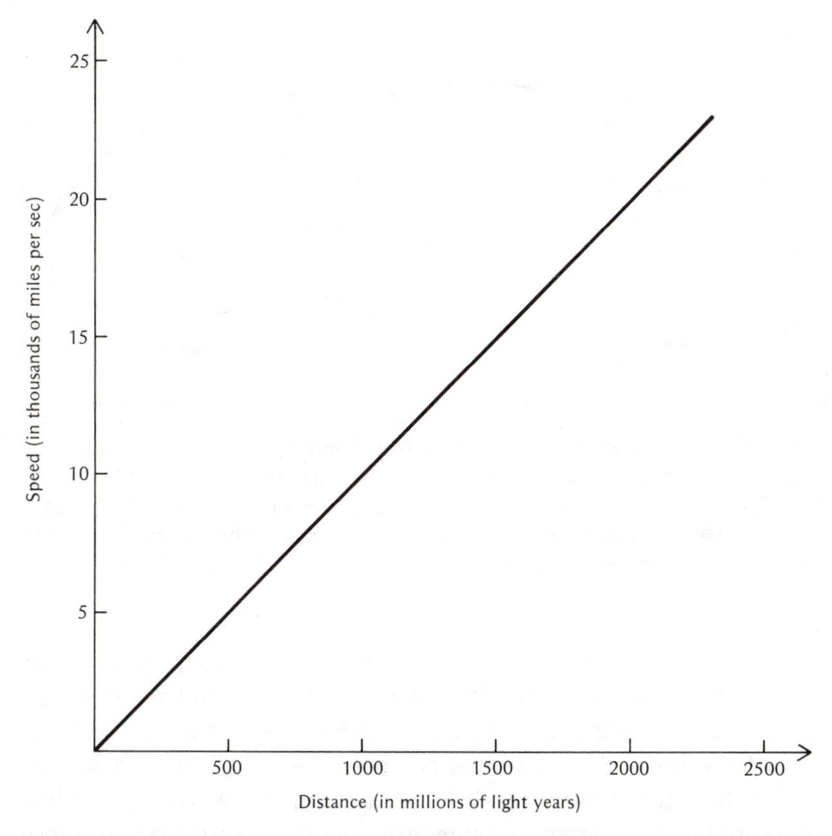

Figure 5-3. The Hubble law. Dr. Hubble discovered that the farther away a galaxy is, the faster it is receding from us. This graph, a plot of the speeds of galaxies against their distances, reveals the direct relationship between these to quantities.

ing an old-fashioned raisin cake. At one o'clock in the afternoon, grandma mixes some yeast and raisins in the dough for her raisin cake and places the loaf on the kitchen table. Initially the loaf is one foot across. Because of the yeast, by two o'clock the loaf has expanded to a diameter of two feet. The "before" and "after" views of our raisin cake are shown in Figure 5-4. Unknown to grandma, however, there is a very intelligent bug sitting on raisin *A*. At one o'clock, this bug measures the distances to raisins *B, C, D,* and *E*. From its measurements it finds the distance shown in the diagram. After an hour's nap, this clever bug repeats its observations at two o'clock. Clearly it finds that the distances from raisin *A* to raisins *B, C, D,* and *E* have doubled. For example, the distance from raisin *A* to raisin *B* has gone from 2 inches to

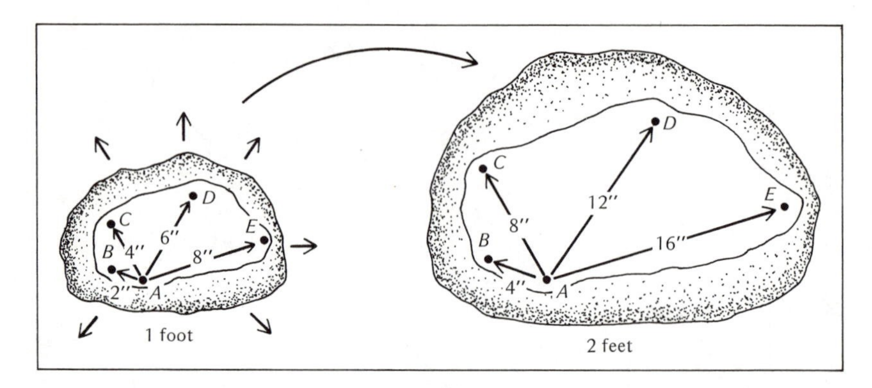

Figure 5-4. The raisin cake. In the "before" picture, we see the cake just after the yeast is mixed in. In the "after" picture, we see the cake one hour later, when the dough has risen. Our intelligent bug is located on raisin *A*. He measures the distances to other raisins before and after the yeast has caused the dough to rise.

4 inches, so the speed of raisin *B* is 2 inches per hour. On the other hand, the distance from raisin *A* to raisin *E* has gone from 8 inches to 16 inches. Consequently, the speed of recession of raisin *E* away from raisin *A* is 8 inches per hour. This bug, therefore, concludes that the nearby raisins are moving away from it slowly, while the more distant raisins are moving away from it more quickly. It is in this fashion that the expansion of the raisin cake is revealed to the bug on raisin *A*.

Clearly, this situation is reminiscent of the Hubble law, which states that nearby galaxies are receding from us slowly, while the more distant galaxies are moving away from us more quickly. By analogy with grandma's raisin cake, we therefore conclude that *the universe is expanding.*

In addition, from the orientation of the straight line in Figure 5-3, we can determine how the recessional velocity is related to distance. The so-called *Hubble constant* tells us how far away you have to be have a certain recessional speed. Recent observational data indicate that the value of the Hubble constant is about 50 kilometers per second per megaparsec, or about 10 miles per second per million light years. In other words, ideally, at 1 million light years from the earth, an object would be receding with a speed of 10 miles per second. At 2 million light years, the recessional speed is 20 miles per second, and so forth.

The Hubble law and the Hubble constant are important tools in astronomy. By measuring the redshift of an extragalactic object, it is possible to infer the distance to that object. For example, if from the redshift of a galaxy we calculate the recessional speed of 10,000 miles per second, we assume the distance to the object to be 1 billion light years.

It should be pointed out that this method of distance determination is by no means iron clad. First of all, the exact value of the Hubble constant is not entirely certain. Measuring the value of the Hubble constant is an extremely difficult problem in modern astronomy. Astronomy books and articles written before 1970 quote values of the Hubble constant that are twice as large as those given here. However, in the mid-1970s, Dr. Alan Sandage at Hale Observatories presented strong evidence that the true value is 10 miles per second per million light years. This physically means that the universe is twice as big and twice as old as we had thought.

The redshift associated with the Hubble law may be called the *cosmological redshift.* As we have seen, it has its origin in the overall expansion of the universe. But there are local effects that may confuse the picture. Consider, for example, the Virgo cluster of galaxies. This cluster in the constellation Virgo consists of hundreds of galaxies, all moving in various directions about their common center of mass. Some are coming toward us, some are moving away, but, on the average, the center of the cluster is receding according to Hubble's law. If we look only at an individual galaxy in this cluster, it is impossible to know offhand how much of its redshift is due to the cosmological effect and how much is due to the simple Doppler effect resulting from the particular motion of this galaxy.

There is, finally, another effect that complicates this subject. One of the famous predictions of Einstein's general theory of relativity is that clocks slow down in a gravitational field. Because atoms radiating light may be thought of as little clocks, the frequency of light emitted by atoms in a gravitational field will be lowered. As a result, the spectrum of this light will be shifted toward the red. Consequently, this phenomenon is known as the *gravitational redshift.*

Thus there are three different possible sources for the redshift of an astronomical object; namely:

1. Doppler effect
2. cosmological redshift
3. gravitational redshift

and it is sometimes very difficult to distinguish one from the other. For example, perhaps the most important advance in astronomy in the 1960s centered around the discovery of quasars. Quasars at first glance look like bluish stars, but their redshifts are truly enormous. If the cosmological interpretation is correct, then quasars are by far the most distant objects ever observed. Indeed, they may be up to 10 billion light years away. This poses many severe problems for the astrophysicist. For example, astronomers know of no physical process that could cause an object to shine so brightly that it could be seen over a distance of billions of light years. To overcome this difficulty, it was proposed that perhaps quasars are nearby, and thus we do not need to trouble ourselves with an inexplicably huge energy output. Perhaps quasars are extremely massive objects, and the redshift comes from the intense gravitational field of the object. As will be seen in a later chapter, calculations demonstrate that such an object is very unstable and will collapse to form a *black hole.*

A third possibility, once proposed by Dr. Halton Arp at Hale Observatories, is that quasars are nearby but traveling at enormous speeds. There is some impressive evidence that the centers of galaxies are the locations of violent activity. Perhaps quasars have been ejected from active galactic nuclei.

Although the mysteries and debates about quasars are far from being resolved, no matter which approach is taken, there is strong motivation for exploring the full implications of Einstein's general theory of relativity. If we take the cosmological approach, then the overall nature of the universe, its shape, its size, and its age become critical questions. If we look to the gravitational redshift, we then are faced with the problem of black holes. Or if we consider active galactic nuclei as a violent source of energy and matter, we find that we must contemplate *white holes,* which are the time reversals of black holes.

In any case, the discovery of quasars has stimulated a great deal of activity among astrophysicists in general relativity. In succeeding chapters, we shall examine some of the results and implications of recent theoretical calculations in relativity theory.

The Black Hole

It is fairly well known that the same physical processes that occur in the hydrogen bomb are also responsible for the sun's energy. In these themonuclear reactions, four hydrogen nuclei (4 protons) are converted into one helium nucleus (2 protons + 2 neutrons) with the accompanying release of energy. To be precise, if we weigh the ingredients and the end-products of this reaction we find that a small amount of matter seems to have disappeared. Four hydrogen atoms weigh slightly more than one helium atom. This missing mass, of course, has been converted into energy according to Einstein's famous equation $E = mc^2$. It is in this way that physicists explain the enormous energy output of the sun and most of the stars we see in the sky. Indeed, at the sun's center, 600 million tons of hydrogen are converted into helium every second!

In about 5 billion years, the major reserves of hydrogen fuel at the sun's center will be exhausted. As is the case with all stars that have reached this stage in their evolution, different nuclear processes take over. Helium is converted into carbon and oxygen. And at still later stages, carbon is converted into heavier elements. Nuclear reactions between heavier elements are far less efficient in producing energy than those involving the lighter elements. Consequently, at some point late in a star's life, all the available sources of nuclear fuel have been used up. The character of a burned-out star then suddenly and violently changes, and the remains of the star assume one of three theoretically possible conditions.

After nuclear fuel has been exhausted, the star begins to contract. There are no longer any sources of energy to hold the star up, and it rapidly decreases in size. Atoms are squeezed closer and closer together, and electrons become disassociated from the nuclei. If the mass of the star in the process of collapsing is less than about 1¼ solar masses, then at some point the pressure from the electrons becomes sufficient to halt the contraction. The electrons act together in a peculiar fashion to hold up the star. Such a star is called a *white dwarf.* White dwarfs are fairly common and have frequently been observed by astronomers. These stars are about the same size as the earth, but because so much matter is confined to such a small region, the density is quite high. One tablespoon of matter from a white dwarf weighs 1000 tons.

If the mass of the collapsing star is slightly greater than 1¼ times the mass of the sun, then the electron pressure is not strong enough to stop continued contraction. The star simply proceeds to become more and more compact. As the pressures in the star increase, electrons collide violently with the nuclei. During such collisions, the negatively charged electrons combine with the positively charged protons to produce neutrons. When, finally, about 90 percent of the star consists of neutrons, the very strong nuclear forces between these particles may be sufficient to stop further contraction and support the star. Such an object, which is only 15 or 20 miles in diameter, is called a *neutron star.* Because so much matter is confined to such a small volume, the densities are huge. One tablespoon of matter from a neutron star would weigh 10 billion tons! With the discovery of pulsars in 1967, we feel that for the first time we are seeing neutron stars. It seems reasonable to suppose that a pulsar consists of a rapidly rotating neutron star with intense magnetic fields.

If the mass of the collapsing, burned-out star is more than

three times that of the sun, then nothing can stop contraction. Neither electron pressure nor nuclear forces between neutrons is sufficient. There are simply no known physical forces that could prevent the star from getting smaller and smaller. As the star becomes more compact and its size continues to decrease, the intensity of the gravitational field increases dramatically. Gravitational force, which normally is the weakest force in nature, now overpowers everything. Because, according to general relativity, gravity is the curving of space-time, space around the star is severely warped. As stellar collapse continues to its inevitable end, this warping becomes so great that space-time folds in over itself, and the star disappears from our universe! What is left is called a *black hole.*

Many stars are far more massive than the sun. Stars of 20, 30, or 40 solar masses are quite common. Up until the mid-1960s, it had been quite fashionable to assume that such stars lost most of their mass after exhausting their nuclear fuel. In this way, they would become white dwarfs, which was the widely accepted final stage of stellar evolution. Neutron stars and black holes had existed on paper since the theoretical work of Dr. Oppenheimer in 1939, but until the discovery of pulsars, few astrophysicists had taken these possibilities very seriously.

With the discovery of quasars and pulsars in the 1960s, this situation changed dramatically. Few astronomers now doubt the existence of neutron stars, and one of the most exciting advances during the 1970s may prove to be the detection of numerous black holes throughout our galaxy and the universe. After all, stars many times as massive as the sun must evolve much more rapidly than the sun. If they were not able to expel most of their mass after exhausting their nuclear fuel, what happened to their remains?

Theoretically, it is quite reasonable to suppose that black holes are very common. The real problem then becomes answering questions such as: What does a black hole look like? How would we recognize a black hole if we saw one? What should we look for in an observational search for black holes? The basic clue to help us in answering these and similar questions lies in the fact that a black hole consists of an extremely intense gravitational field. In the 1960s, many astrophysicists, therefore, spent a great deal of effort in determining the behavior of light rays (photons) and particles in the highly warped space-time surrounding a black hole.

Suppose that we have before us one of these black holes. Imagine that light rays are trying to pass by this black hole at various distances. Because space-time is highly curved by the gravitational field of the black hole, these light rays will be bent from their classical straight-line paths. Light rays passing the black hole at large distances will be deflected only slightly. Light rays passing nearer the black hole will be more severely deflected. In fact, it is possible to aim a beam of light toward a black hole so that the light will go into a circular orbit. This situation is seen most clearly in Figure 6-1. This circular orbit is called the *photon circle* or the *photon sphere.* Every star in the visible universe contributes light to the photon sphere surrounding a collapsed massive star. This circular orbit is, however, highly unstable. If a photon deviates even slightly from this very precise photon sphere, the light will rapidly spiral inward or outward away from the circular orbit.

Finally, beams of light that are aimed still more directly at the black hole will ultimately be sucked into the black hole. In fact, all photons that come nearer to the black hole than the photon sphere are forced to spiral in and collide with the black hole.

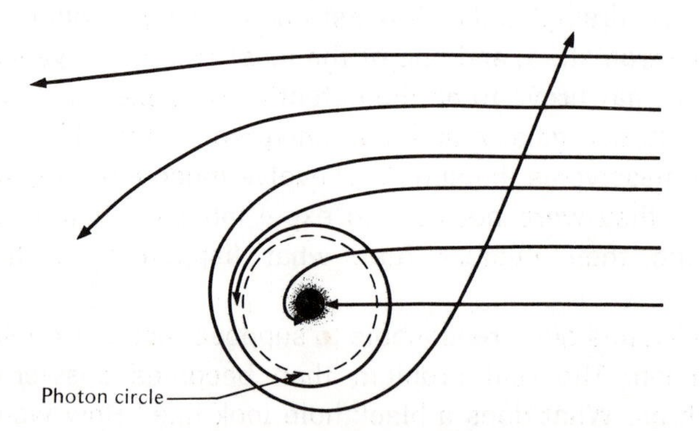

Figure 6-1. Deflection of light by a black hole. Beams of light passing near a black hole will be deflected through large angles. A beam of light approaching a black hole at precisely the right distance will be captured in the *photon circle.* Beams of light aimed more directly at the black hole will be sucked into the black hole.

In this regard, it should be pointed out that a beam of photons or a light ray defines a straight line. It is impossible to draw a line straighter than the beam from a laser, for example. Therefore, when we speak of circles and spirals followed by photons, we are really talking about straight lines in a warped four-dimensional space-time.

To understand the nature of black holes more fully, imagine standing on the surface of a massive, burned-out, collapsing star. The star rapidly decreases in size, and the gravitational field of the star gets stronger and stronger. (Actually, long before we get to the conditions where strange effects due to relativity begin occurring, the gravitational stresses are so great that human life is quite impossible. We shall ignore this minor problem in our discussion.) Everything proceeds as expected until the collapsing star has fallen inside its own photon sphere. When the star has shrunk to a smaller size than its photon sphere, strange things begin to happen. Imagine that we have a powerful searchlight on the surface of the star. If the searchlight is pointed directly upward, the beam of light escapes from the star and goes straight off into space. If we tip the searchlight slightly away from the vertical, the beam of light will be bent slightly by the curvature of space-time. As we tip the searchlight over still further, the deflection of the light becomes greater. Finally, at some angle the deflection is so great that the beam of light can no longer escape from the star. The light goes up but is bent so much that it comes back down and strikes the surface of the star.

We can define an imaginary cone, the *exit cone,* centered about the vertical direction, as shown in Figure 6-2. All beams of light emitted from the surface of the star inside this cone will escape from the star. All beams of light emitted from the star outside this cone cannot escape; the photons go up and the photons come down. Photons emitted at an angle from the vertical equal to the half-angle of the cone neither escape nor fall back; they go into circular orbit about the star in the photon sphere.

As the star continues to collapse further, the warping of space-time becomes more and more severe. From the viewpoint of an observer on the surface of the star, the exit cone becomes increasingly narrow. With continued collapse, you must aim the searchlight closer and closer to the vertical in order for the light to escape from the star. Finally, at a very important stage in

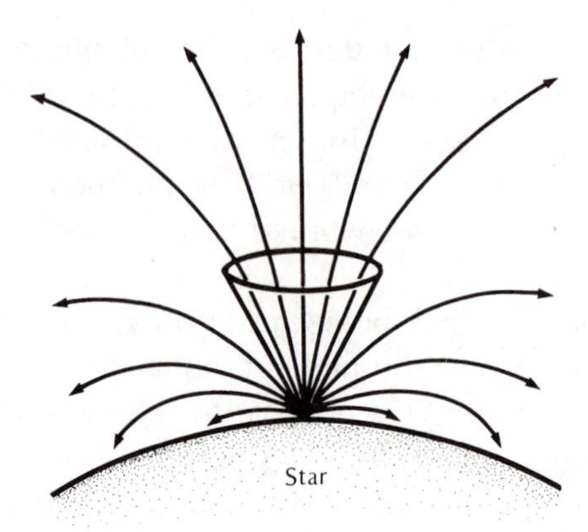

Figure 6-2. The exit cone. On the surface of a collapsing star, we can define an imaginary *exit cone.* All light rays leaving the star at angles inside this cone will escape from the star. All light rays emitted at angles outside the exit cone cannot escape from the star. Such rays are bent so severely that they simply go up and come back down.

relativistic gravitational collapse, the warping of space-time becomes so great that the exit cone completely closes up! In other words, no matter what direction you point the searchlight, the light cannot escape from the star. In fact, the gravitational field is so strong that absolutely nothing can ever again leave the star; the escape velocity is greater than the speed of light. The moment that the exit cone completely closes up, we say that we have crossed the *event horizon.* In no way can we communicate with the universe outside. In a very real sense, for someone on the outside watching this collapsing star, the moment we irretrievably cross the event horizon we have disappeared from the universe! Graphs showing the diameter of the photon sphere and the event horizon as a function of the mass of the black hole are shown in Figures 6-3 and 6-4.

After passing through the event horizon, the star continues to collapse still further. Pressures and stresses due to the incredibly intense gravitational field finally become infinitely great. Everything of which the star was composed is crushed beyond recognition. When the remains of the entire star are crushed into

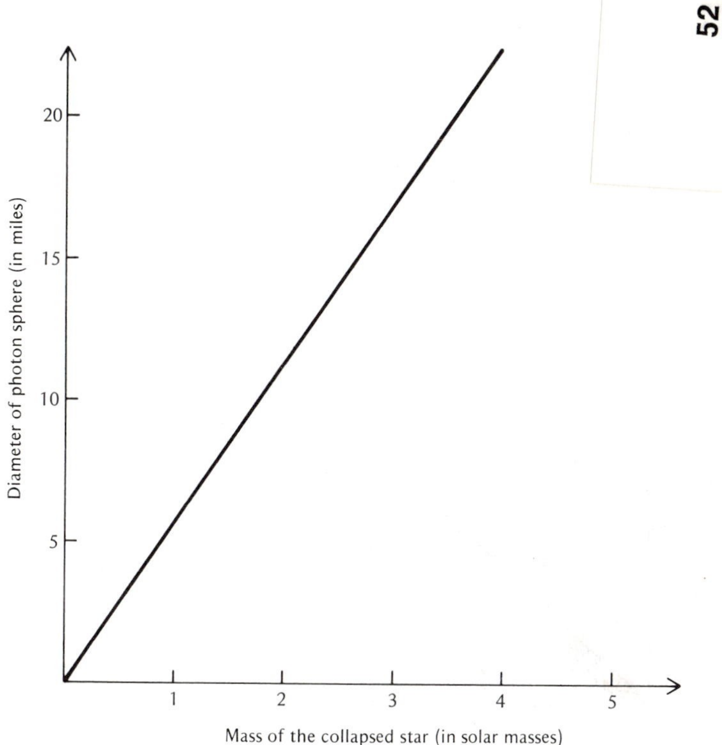

Figure 6-3. Diameter of the photon sphere. This graph shows how the diameter of the photon sphere varies as a function of the mass of the black hole (Schwarzschild solution). For example, the black hole of a collapsed star three times as massive as the sun is surrounded by a photon sphere about 17 miles in diameter.

zero volume, the pressure and density are infinite. At this point we have reached the so-called *singularity*.

A very important effect that should not be overlooked in this treatment is the gravitational redshift. Time slows down in a gravitational field. To see what this means, suppose we watch as a foolhardy scientist plunges into a black hole. As he gets closer and closer to the black hole, we observe that his clocks, (that is, wristwatches, his pulse, or anything else used to measure time) are slowing down. In fact, we claim that his clocks will stop completely when he reaches the event horizon. Of course, we never see this scientist reach the event horizon; as his clocks appear to slow down, he also appears to travel slower and slower. We claim that it would take an infinite amount of time for this

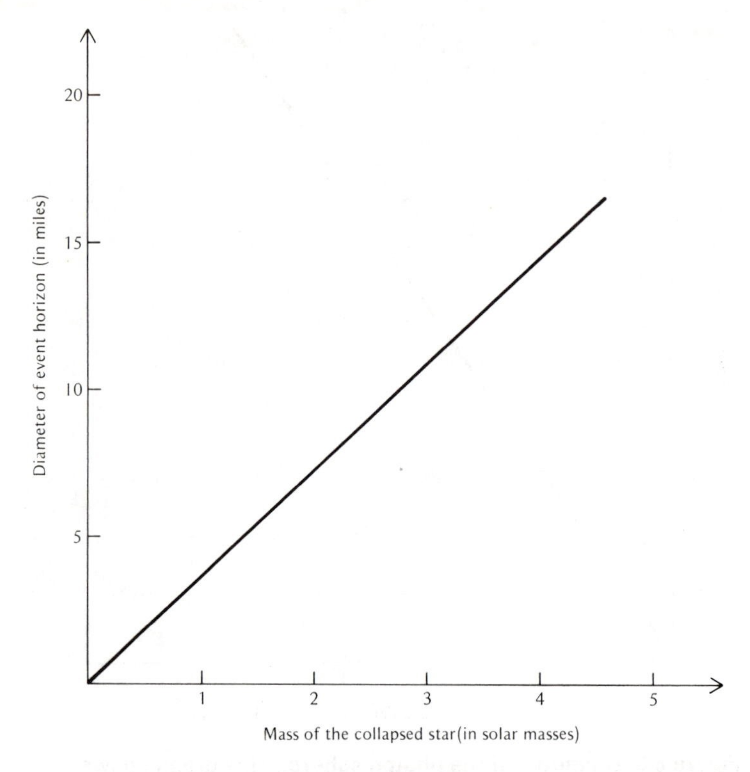

Figure 6-4. Diameter of the event horizon. This graph shows how the diameter of the event horizon varies as a function of the mass of the black hole (Schwarzschild solution). For example, the black hole of a collapsed star three times as massive as the sun is surrounded by an event horizon about 11 miles in diameter.

scientist to cross the event horizon. The foolhardy scientist says that this is all wrong. He looks at his wristwatch and does not notice anything unusual (of course, his heartbeat, thinking processes, and everything else has slowed down by exactly the same amount). And indeed, this scientist finds that, according to his wristwatch, he actually does cross the event horizon in a finite amount of time.

Observationally, this means that when looking at light from the surface of a collapsing star, two simultaneous but separate, effects are seen. First of all, the star becomes dimmer and dimmer because it is increasingly more difficult for light to escape from the star. Photons have to be emitted at angles inside the exit cone or they just do not escape. Second, because atoms are like little clocks, as light is emitted by atoms on the star, the frequency of the radiation decreases as the collapse proceeds. The entire spectrum is shifted more and more toward the red. In other words, as we watch a collapsing star, it should appear increasingly dimmer and redder as time goes on.

It should be emphasized that this treatment of the black hole is very idealized. It is really valid only for a perfectly spherical, nonrotating star. The solution to Einstein's field equations for a static, spherically symmetric black hole was first given by Karl Schwarzschild in 1916. This solution is therefore called the *Schwarzchild solution.* Of course, all stars rotate. As they collapse, their rate of rotation will speed up, just as an iceskater speeds up by pulling in her arms as she does a piroutte. In addition, stars are not perfectly spherical, and as collapse proceeds, the star could easily take on shapes that are rather odd. One of the basic thrusts in theoretical astrophysics during the 1970s deals with how these deviations affect the formation of black holes. In these investigations, many new powerful mathematical tools will be brought to bear on the problem. A pioneer in this field is Dr. Roger Penrose of the University of London. He has shown that if a star collapses within its own event horizon, no deformations whatsoever can prevent the formation of a singularity.

In the next chapter, we shall see that if rotation is included in the treatment of black holes, even after crossing the event horizon, you are *not* forced to collide with the singularity. This has fascinating implications for space travel to other universes.

White Holes and Wormholes

In 1916, only months after Dr. Einstein published his general theory of relativity, the German astronomer Karl Schwarzschild discovered the first "rigorous" solution to the field equations. This famous, but simple, solution describes how space-time is warped by the gravitational field of a massive collapsed star or black hole. The Schwarzschild solution assumes that the collapsed star is perfectly spherical and not rotating. Far from the star, the gravitational field is virtually identical with Newton's inverse-square law of gravity. Near the star, however, a number of strange things happen that are entirely foreign to classical, prerelativistic theory.

As we approach a black hole, described by the Schwarzschild solution and shown schematically in Figure 7-1, we first come to the *photon sphere,* which consists of a thin shell of light rays or photons orbiting the collapsed star. As we fall further in toward the star, we pass the *event horizon.* Once we pass through the event horizon, it becomes impossible to communicate with the outside universe. We have, in essence, disappeared from the universe. Finally, at the very center, we reach the *singularity.* Here we encounter infinite pressures due to an inconceivably intense gravitational field. This would not make a very pleasant spaceflight; even the nuclear particles out of which our spaceship was composed would ultimately be crushed beyond all recognition.

Because the Schwarzschild solution is mathematically the simplest of the static, spherically symmetric, nonrotating solutions to Einstein's field equations, it has been studied in detail even up to the present day. In the mid-1930s, Drs. Einstein and Rosen examined the Schwarzschild solution very carefully and found something truly remarkable.

To appreciate the work of Einstein and Rosen, consider a star that is undergoing gravitational collapse. All the nuclear fuel has been used up, and the star is becoming smaller and smaller. As the star becomes more compact, the strength of the gravitational field increases, which means that space-time becomes highly warped. If we could photograph this curved space, it would look something like that shown in Figure 7-2. Far from the star, space-time is essentially flat, while near the star, it is highly curved. The body of the star is indicated by the shaded region.

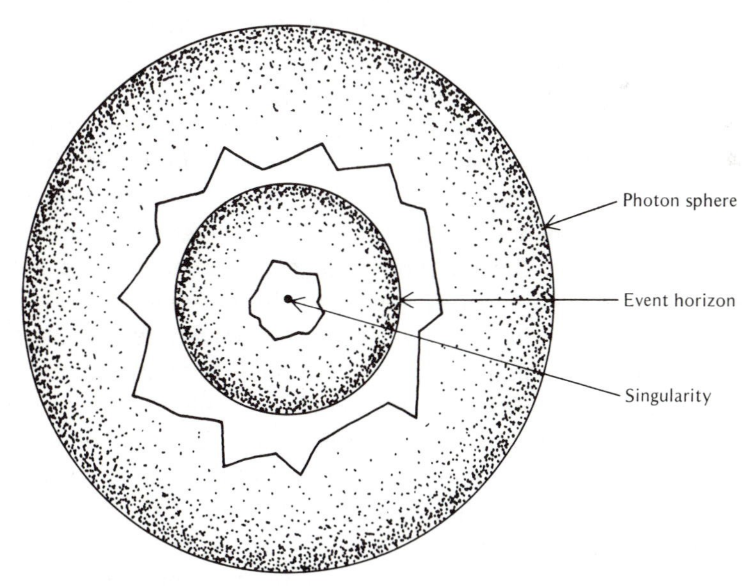

Figure 7-1. The black hole. A schematic diagram of a black hole showing the photon sphere, the event horizon, and the singularity.

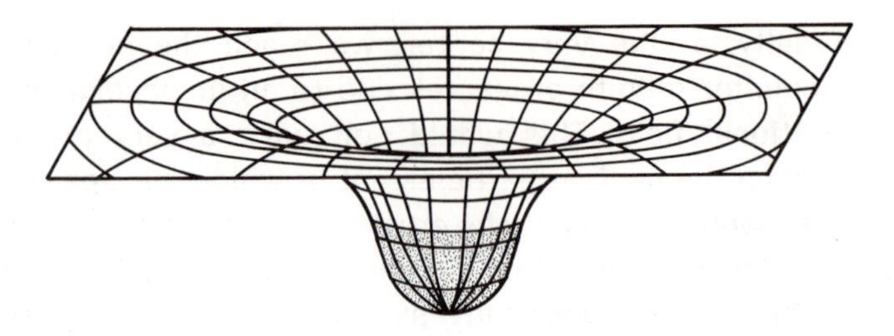

Figure 7-2. Warped space-time. Space near a collapsing star is curved by the gravitational field of the star in the same way a rubber sheet would be curved by a brick or some other heavy object placed on it. The rest of the universe, which may be assumed to be essentially flat, is far away from the collapsing star.

Figure 7-2 is called an *embedding diagram.* Such diagrams are obtained by "slicing" through warped space-time. The resulting spacelike slice reveals the curvature of space, just as slicing through a layer cake reveals how the layers of cake and icing are arranged. In drawing embedding diagrams, two of the four dimensions of space-time are suppressed. This allows us to visualize more easily the geometry of curved space associated with an intense gravitational field.

If we follow the continued collapse of a star, the curvature of space show in Figure 7-2 becomes more and more severe. The eventual fate of this curved space was first described by Einstein and Rosen. Surprisingly, they found that the embedding diagram eventually opens up into a second universe. After passing the event horizon, the curvature of space becomes less severe and emerges into a second asymptotically flat universe. We shall later see that spaceflight between the two universes is impossible because to take such a trip you would have to travel much faster than the speed of light. What we have here is simply the geometry of space around a black hole. Essentially, Einstein and Rosen found that the Schwarzschild solution actually consists of *two* solutions to the field equations back to back. One piece of the complete solution connects our universe to the event horizon, and the other piece continues on into another universe. This geometry is shown in Figure 7-3. The resulting geometrical figure is called a *wormhole,* or the *Einstein-Rosen bridge.*

At this point we should pause to understand thoroughly what

is meant by Figure 7-3. This is a two-dimensional drawing of a four-dimensional process. Consider the collapse of an idealized, spherical, nonrotating star as seen by an outside observer. The spherical surface of the star gets smaller and smaller as time goes on. We may imagine the star as a ball being squeezed into the wormhole. As the ball contracts, it can slip deeper and deeper down the throat of the Einstein-Rosen bridge. The smaller the ball, the farther it fits down the wormhole. According to the general theory of relativity, however, time slows down as the strength of the gravitational field increases. Consequently, as seen by an outside observer, the rate of contraction seems to slow down. Because, according to the distant observer, time stops altogether at the event horizon, the outermost atoms above the contracting star never reach the event horizon at the middle of the Einstein-Rosen bridge.

The first interpretation of the Einstein-Rosen bridge that comes to mind is that we have connected two separate, distinct universes. This is not the only interpretation, however. The Einstein-Rosen bridge could be connecting two very distant points in our own universe. How this might happen is shown in Figure 7-4. All this says is that the upper sheet may be connected to the lower sheet. If you prefer to think of the universe as essentially flat, Figure 7-4 can be unfolded to obtain Figure 7-5. These two drawings are entirely equivalent.

The careful reader will have noticed that the Einstein-Rosen bridge does not tell the whole story. Admittedly, from the viewpoint of someone who stands off at a distance and watches the

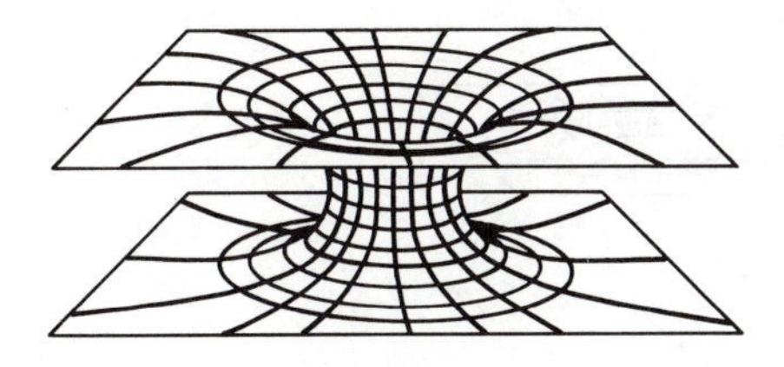

Figure 7-3. The wormhole. This diagram shows the geometry of space-time about a nonrotating black hole. The wormhole, or Einstein-Rosen bridge, connects two separate flat universes. The event horizon is at the narrowest section of the wormhole.

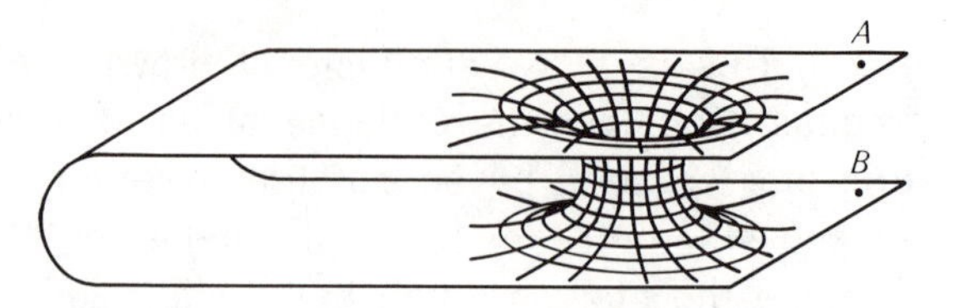

Figure 7-4. The wormhole. This diagram shows the Einstein-Rosen bridge connecting two distant points (*A* and *B*) in our own universe.

gravitational collapse of a star, the outer layers of the star never managed to reach the event horizon. Such a process would seem to take an infinite amount of time. But according to the wristwatch of a scientist standing on the surface of the collapsing star, he falls through the event horizon in a comparatively short time. The star then continues to contract and finally reaches the singularity. But the singularity is not shown in the Einstein-Rosen bridge. The Einstein-Rosen bridge tells us the geometry of a black hole as seen from the *outside.* An outside observer cannot actually see the singularity, and consequently it does not appear in these drawings.

This does not prevent us from asking about the nature of space-time between the event horizon and the singularity. Because the unexpected double-universe nature of a black hole was hidden in the Schwarzschild solution, perhaps all we have to do is delve a little further. Perhaps there is still more to be learned from the Schwarzschild solution.

In 1960, Drs. M. D. Kruskal and G. Szekeres independently

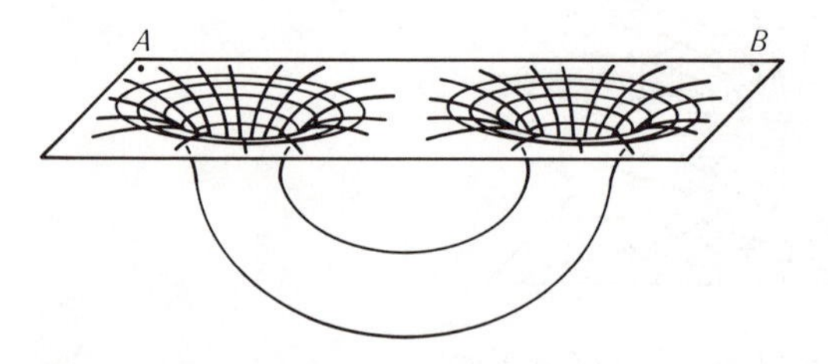

Figure 7-5. The wormhole. This is an unfolded version of the previous diagram. Contrary to what the diagram seems to suggest, the distance directly from point *A* to point *B* may be much longer than the distance through the wormhole connecting these two points.

published their discovery of what is called the *maximal extension of the Schwarzschild metric,* which fully described the nature of space-time all the way down to the singularity. Using a set of very clever coordinate transformations, Drs. Kruskal and Szekeres were able to unlock mathematically many of the hidden treasures contained in the Schwarzschild solution. As in the case of the Einstein-Rosen bridge, we shall present the results of Kruskal and Szekeres in the form of a diagram. In order to prepare for this diagram, we must first clearly understand and appreciate the nature of a mathematical process known as *mapping.*

Consider Figure 7-6a. If asked "What are you looking at in Figure 7-6a?" most people would promptly reply, "The earth." Of course, they are completely wrong. Figure 7-6a does not look at all like the earth. The diagram is flat, the earth is not. The diagram has four sharp corners, the earth does not. The north and south poles are straight lines at the top and bottom of the diagram, and the shapes of many of the continents are severely distorted. How could any intelligent person possibly confuse that weird diagram in Figure 7-6a with the earth?

There are two ways to resolve this dilemma. First of all, ever since you were a small child, people have been telling you that a diagram such as Figure 7-6a is a picture of the earth. More importantly, however, you understand the process of mapping. In order to make Figure 7-6a really look like the earth, you have to cut out the diagram with a pair of scissors, fold it around a ball, and glue it down, as shown in Figure 7-6b. But if you want a picture or map of the earth, it is not always convenient to carry around a globe. Because you understand the process of map-making, you find that you are willing to sacrifice a certain amount of realism in order to have a flat piece of paper that you can fold up and put into your pocket. The price you have to pay for this convenience is that you are now faced with a variety of distortions. For example, the north and south poles, which are points on the real earth, become straight lines along the top and bottom of the map. The equator, which is really a circle, also becomes a straight line. In other words, unusual geometrical things happen, but you are not overly disturbed because you understand what is going on.

The best way to present the work of Drs. Kruskal and Szekeres is in the form of a very special map called a *Penrose diagram.*

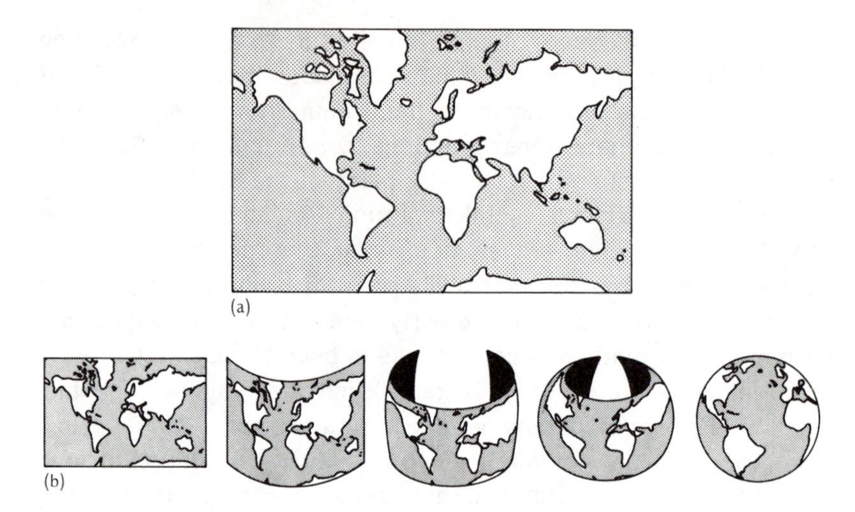

Figure 7-6. Map of the earth. Here we see how the spherical surface of the earth can be mapped onto a flat plane. Although a flat map may be more convenient, many unusual geometrical distortions enter.

As in the case of maps of the earth, we can expect a number of very strange distortions in this diagram of the space-time around a black hole. But these distortions will not bother us too much, not only because we understand what is going on but also because we shall uncover a great deal of information about the nature of a black hole.

The best way to set up a Penrose diagram is on a graph on which you plot time vertically and distance horizontally. This approach is shown schematically in Figure 7-7. We shall in addition require that, in such a diagram, light rays travel along 45° lines. This may be easily accomplished by properly scaling the time and distance axes: For every second measured in the vertical direction, 186,000 miles is measured in the horizontal direction. If we now imagine an astronaut traveling around in this space-time diagram, we see that there are certain directions in which he can and cannot move. For example, he can easily move from point *A* to point *B*. During such a trip, a great deal of time passes, but not much distance is covered. Such a path is called *timelike*. If he wants to go from point *C* to point *D*, however, he must travel at the speed of light. The line connecting points *C* and *D* is at 45°, and such a path is called *lightlike*. Finally, he can never travel from

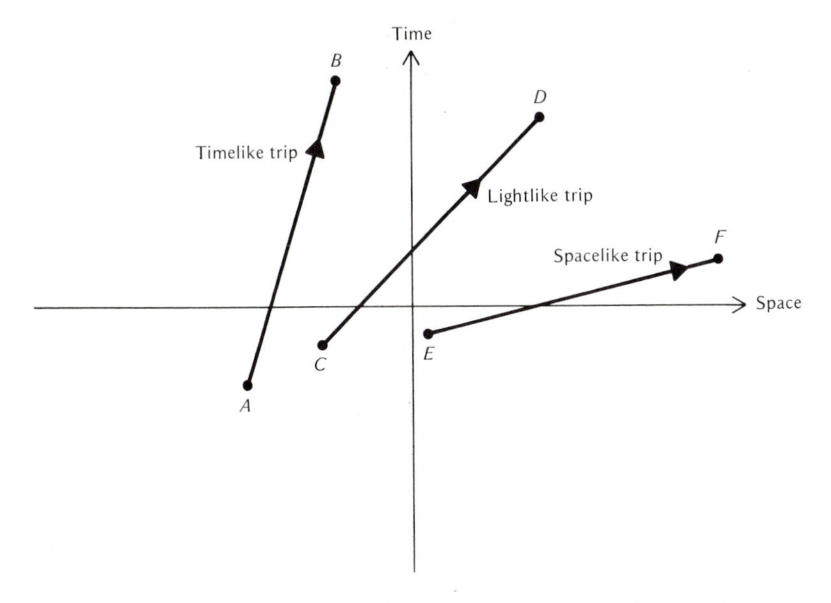

Figure 7-7. Paths in space-time. The path joining points *A* and *B* is timelike. The path joining points *C* and *D* is lightlike. The path joining points *E* and *F* is spacelike. Because you can never travel faster than the speed of light, only timelike trips are permissible.

point *E* to point *F*. A great deal of space must be covered in a short period of time; his speed would have to be greater than the velocity of light. Such a path is called *spacelike.* It is, therefore, important to realize that all matter (an astronaut, the surface of a collapsing star, and so on) can travel only along timelike paths. These permitted paths in this diagram lie at angles less than 45° with respect to the vertical. It is totally impossible to travel along a path inclined by more than 45° from the vertical.

We are now in a position to examine the Penrose diagram of a nonrotating, spherical black hole shown in Figure 7-8. This diagram is set up so that light rays travel along 45° lines. The first distortion that you notice is that our entire universe is lined up very near the left-hand side of the figure. The "other universe," which is the same as the second sheet of the Einstein-Rosen bridge, is lined up near the right-hand side of the figure. The singularity, shown as wavy lines, bounds the top and bottom of the diagram. Finally, we see that the event horizon, shown as two lines intersecting at the center of the figure, divides all space-time into two distinct types of regions. One type of region (I) is

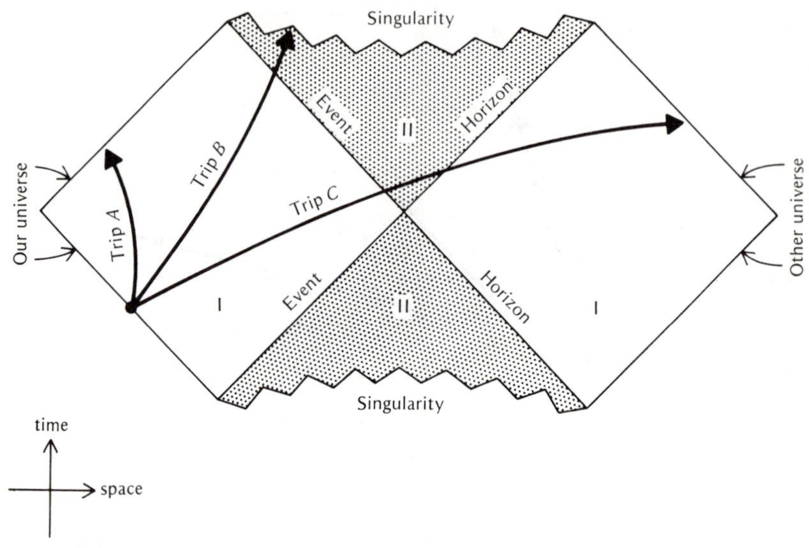

Figure 7-8. The Penrose diagram of the Schwarzschild solution. Our universe is lined up very near the left side of the diagram. The other universe (the lower sheet of the Einstein-Rosen bridge) is lined up near the right side of the diagram. The singularity bounds the upper and lower parts of the diagram, while the event horizon crosses through the center. Region I consists of the areas between the event horizon and the outside universes. Region II comprises the areas between the event horizon and the singularity. Trips *A* and *B* are permtited. Trip *C* is spacelike and therefore violates the laws of physics.

everything outside the event horizon. There are two such areas: one between us and the event horizon and one between the event horizon and the other universe. Obviously there are two regions of the second type (II) between the event horizon and the singularity.

Remembering that astronauts (or any physical object) must travel along timelike paths, we can now understand the reasons behind many of the statements that have been made in this and the preceding chapter. Consider a variety of possible trips that an astronaut might try to make. On trip *A*, the astronaut flies past the black hole. This is a safe, permissible trip; the entire trip is timelike. Trip *B* is also permissible, but this time the astronaut foolishly passes through the event horizon. Two things are immediately clear from the Penrose diagram. First of all, because in this diagram the event horizon is inclined at 45°, once the astronaut is inside region II, it is impossible for him to get out. In order to get out, his escape path would have to become spacelike, which means that he would have to travel faster than the speed of light. Because this is impossible, the inevitable fate of the astronaut is to be crushed when he hits the singularity. In addition, once inside region II, between the event horizon and the singularity, it is impossible for the astronaut to communicate with the outside world. All the light rays or radio waves that he sends out must travel along lightlike trajectories. All such trajectories must ultimately strike the singularity and can never escape from region II. Finally, trip *C* is totally impossible. At various points along this path, the trajectory is inclined by more than 45° to the vertical. In other words, there are spacelike portions of the trip, and the astronaut would have to travel faster than light. All imaginable trips from our universe to the "other universe" must have some portion that is spacelike. Therefore, using the wormhole as a means of space travel to other universes is completely futile.

On numerous occasions in this treatment of the Schwarzschild solution, we have emphasized that the black hole was not rotating. This is a severe drawback in terms of physical reality. Everywhere we turn we observe rotation: planets rotate, stars rotate, galaxies rotate, and perhaps the entire universe is rotating. Obviously the omission of rotation is an unfortunate and severe limitation. Up until very recently, however, astrophysicists simply did not know how to include rotation into their calculations. There were no known solutions to Einstein's field equations that properly took rotation into account.

In 1963, a major breakthrough occurred in relativity theory when Dr. R. P. Kerr published his discovery of a solution to the field equations that properly incorporated rotation. This so-called *Kerr solution* contains even more surprises than the old, nonrotating Schwarzschild solution. First of all, there are *two* event horizons in the Kerr solution instead of one. In essence, as soon as you start spinning the Schwarzschild solution, a second (inner) event horizon appears. In addition, the entire character and properties of the singularity change dramatically.

As you might by now expect, the best way to display Dr. Kerr's discovery would be through a *Penrose diagram of the Kerr solution.* Such a diagram was first discovered in 1967, with the work of Drs. Boyer and Lindquist, and is shown in Figure 7-9. As

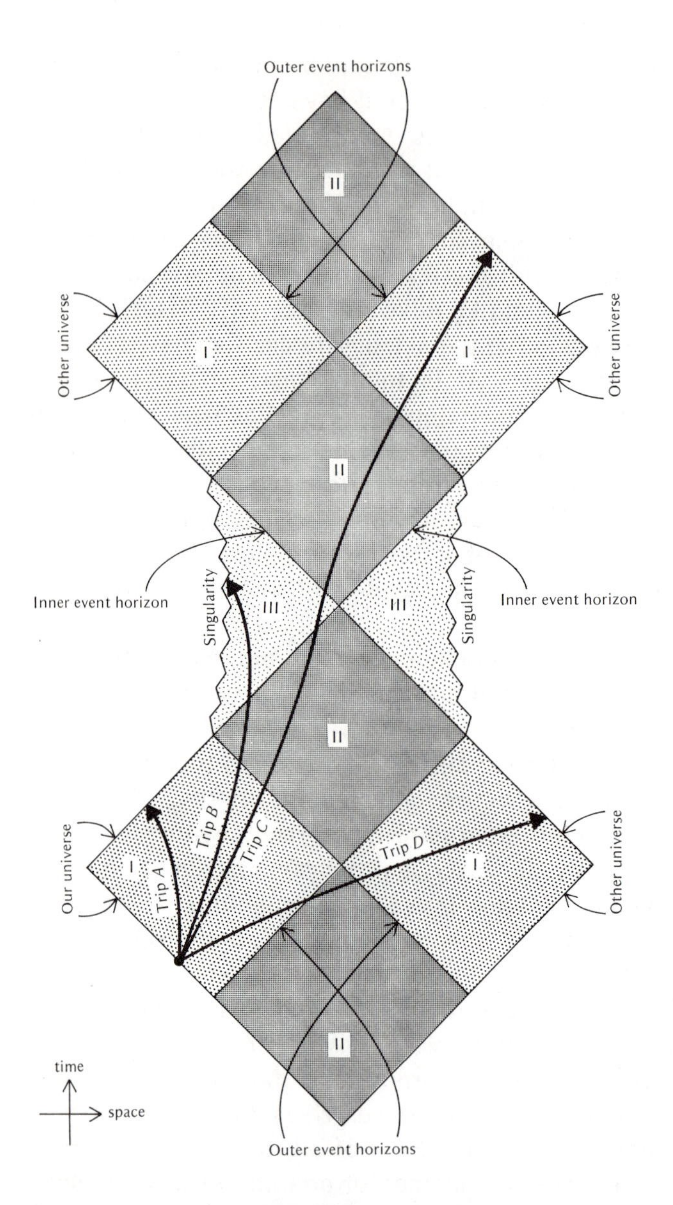

Figure 7-9. The Penrose diagram of the Kerr solution. The geometry of a rotating black hole connects many universes. Our universe is shown lined up in the lower left portion of the diagram. Three other universes are also shown. There are now two event horizons, the inner and the outer. The region I areas are located between the outer event horizon and the outside universes. Region II are those areas between the two event horizons. Region III consists of those areas between the inner event horizon and the singularity. Trips *A*, *B*, and *C* are permitted, but trip *D* is not.

before, light rays travel along 45° lines. Although this diagram looks unbelievably complicated, there are a few very interesting surprises in store.

It is clear from the diagram that the two event horizons that crisscross down the center of the figure divide all space-time up into three distinct types of regions. The first type of region (1) is all those areas between outside universes (ours or someone else's) and the first event horizon. The second type of region (II) is all those areas between the first and second event horizon. And, finally, the third type of region (III) is those areas between the second event horizon and the singularity.

The first thing we notice in looking at the Penrose diagram of the Kerr solution is that there are a variety of different universes. Although only four universes (ours and three others) are shown in Figure 7-9, this diagram repeats itself indefinitely over and over again in the vertical direction. Figure 7-9, is therefore, only one small section of an infinitely long strip of diagrams on which the same pattern is constantly repeated. Consequently, we see that a rotating black hole may be thought of as forming a bridge that connects an infinite number of otherwise separate universes.

The second thing we notice about Figure 7-9 is that the singularity (indicated by wavy lines) is located vertically in this diagram, while in the Schwarzschild description the orientation is horizontal. This reflects a fundamental change in the character of this region of space-time. In the Schwarzschild solution we have a *three-dimensional spacelike singularity,* while in the Kerr solution we are confronted with a *two-dimensional timelike singularity.* This has a number of important and signficant implications.

Consider an astronaut who undertakes a variety of space-flights in the vicinity of a rotating black hole. We recall, of course, that only timelike paths are permissible. On path *A* the astronaut leaves the earth but does not get very close to the black hole. He ends up somewhere else in our own universe. On path *B* the astronaut again leaves the earth, but this time he heads directly for the rotating black hole. He passes through the first event horizon and then through the second event horizon. Unfortunately this astronaut makes the fatal mistake of heading directly for the singularity. In approaching the singularity in the equatorial plane of the hole (i.e., perpendicular to the hole's axis of rotation), he is ultimately crushed by infinite pressures.

The tragic end of this ill-fated astronaut was totally unnecessary. One of the surprising properties of the Kerr solution is that if you pass through the event horizons, you are *not* doomed to collide with the singularity. For example, the astronaut could have chosen path *C*. In this case, after passing through regions II and III, he would leave the event horizons behind and emerge in another universe! Because there are innumerable possible flight paths similar to path *C*, it is clear that the rotating black hole as described by the Kerr solution affords us with a means of space travel between different universes.

At this stage of the discussion, we hardly need to pause and remark that a trip along path *D* is impossible. This path is inclined by more than 45° with respect to the vertical and is therefore spacelike. To undertake such a trip, you would have to be able to travel faster than light, which violates many known laws of physics.

It was mentioned that the diagram shown in Figure 7-9 is really one section of a long strip on which the same pattern is repeated over and over. The first possible interpretation, therefore, is that the geometry of a rotating black hole connects an infinite number of universes. To arrive at other possibilities, however, reconsider the discussion of the map of the earth. In order to make the map of the earth (Figure 7-6a) really look like the earth, the first thing we had to do was cut out the picture, fold it around the shape of a cylinder, and glue the vertical seam together. So also with the Penrose diagram of the Kerr solution; we could fold it around in the shape of a cylinder, as shown in Figure 7-10. This alternate interpretation presents the possibility that an astronaut could leave our universe, travel along a timelike path all the way around the cylinder, and return to our own universe. By pointing his spacecraft in the appropriate direction, he can return to almost any point in space-time in our universe that he chooses. For example, he could come back to the earth a billion years ago or a billion years in the future! This is a *time machine.* Indeed, the astronaut could come back to the earth five minutes before he left and meet himself! Such a possibility violates the *law of causality,* and if such trips are feasible, then physical reality is irrational at a very fundamental level. This state of affairs is highly objectionable to most scientists. It is therefore generally believed that the *full* space-time geometry of the pure Kerr solution (as

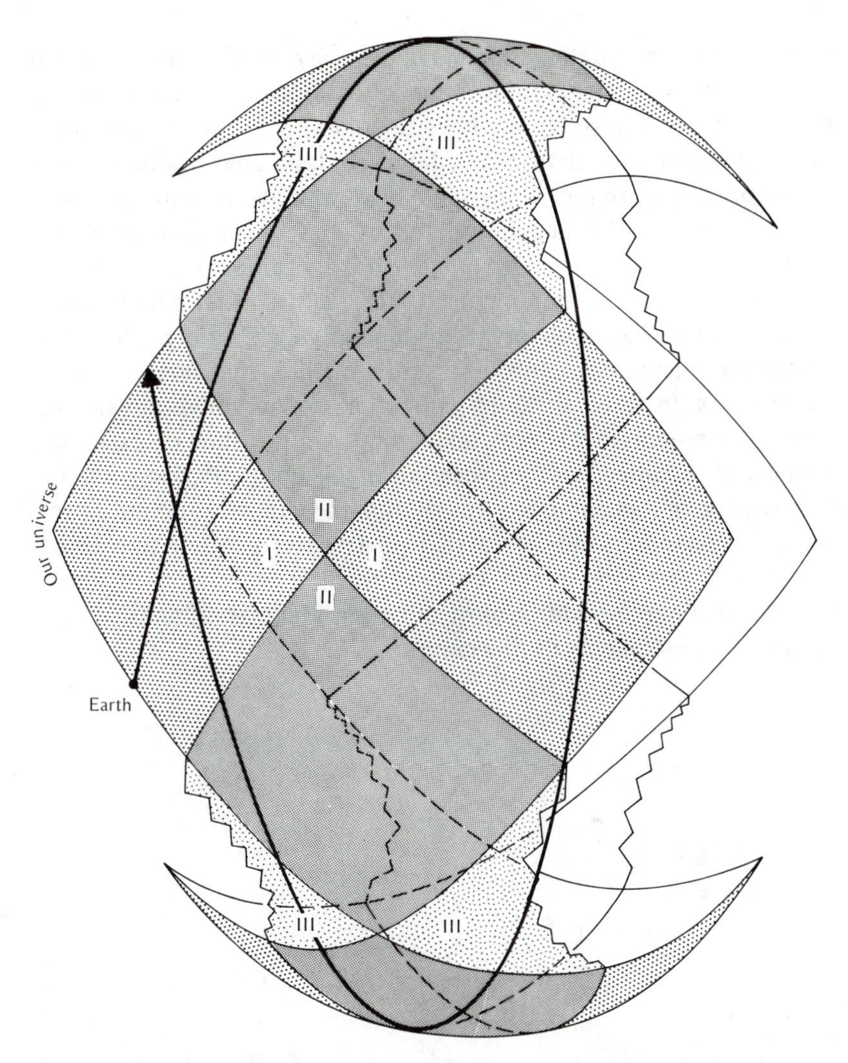

Figure 7-10. The folded Kerr solution. The Penrose diagram of the Kerr solution may be folded in the shape of a cylinder, as shown. This interpretation allows us to travel through a rotating wormhole and return to our own universe, perhaps sometime in the distant past! Clearly this violates causality.

shown in a Penrose diagram) is so idealized that it no longer properly describes physical reality.

Beginning in the late 1960s, reliable observations of the skies in all wavelengths began to pour in from many different observatories and universities. For the first time, astronomers could begin to piece together data ranging all the way from x-rays to radio waves. Although many interesting questions were answered by such observations, scientists have been presented with great quantities of additional data that blatantly defy any traditional explanation. For example, many galaxies and quasars emit far more energy than can be accounted for by the usual thermonuclear reactions. In addition, Dr. Arp at Hale Observatories has discovered a variety of galaxies that appear to be ejecting other galaxies and even quasars. These observations of active galactic nuclei have prompted many astrophysicists to consider the possibility of *white holes.* In spite of the objections of conservative physicists, perhaps the centers of these exploding and peculiar galaxies are actually the "other end" of rotating black holes. Perhaps matter and energy are gushing up out of the throats of rotating wormholes that connect our universe to a variety of other universes. A detailed theoretical and observational examination of these possibilities may provide one of the most fascinating and exciting advances in astronomy of this century.

Discovering Black Holes

As you can imagine, the task of discovering real black holes in the sky poses some monumental problems for the astronomer. In a black hole, gravity is so intense that nothing—not even light—can escape. At first glance, therefore, looking for black holes might seem to be a hopeless quest.

Searching for black holes would be hopeless if it were not for the fact that many stars occur in *binary systems.* Indeed, almost half of the stars seen in the night sky are *binary stars.* A binary system is one in which two stars orbit about their common *center of mass* just as the earth and moon are in orbit about each other. Realizing that there are numerous binary stars in the sky, two Soviet astrophysicists, Drs. Ya. B. Zel'dovich and O. Kh. Guseynov, proposed a method of detecting black holes.

Occasionally astronomers find binary systems in which only one star is visible. It is usually believed that in such cases the second star is simply too small and dim to be observed. Astronomers are nevertheless sure of the fact that a binary system has been discovered due to the characteristic behavior of the spectrum of the visible star. Such systems are called *single-line spectroscopic binaries* because the spectral lines of the visible star alternately shift back and forth as the star orbits its unseen companion.

In 1964, Zel'dovich and Guseynov proposed a careful reexamination of the orbits of all known single-line spectroscopic binaries. Usually the most massive stars are also the brightest stars. If a binary were discovered whose orbit indicated that the unseen companion was very massive, then this massive, invisible star would be a good black-hole candidate. Unfortunately, a careful search through catalogs of binary stars failed to produce any conclusive results. The invisibility of unseen companions could always be explained without resorting to black holes. A similar investigation by Drs. K. S. Thorne and V. Trimble at Caltech in 1968 was equally inconclusive.

The next major step towards the discovery of black holes occurred in the early 1970s. On December 12, 1970, a satellite named *Uhuru* (the Swahili word for freedom) was launched into equatorial earth orbit from a platform in the ocean off the coast of Kenya. The purpose of this satellite was to make x-ray observations of the sky. Equipped with two-x-ray telescopes, the spacecraft scanned the sky sending signals back to earth each time an x-ray source was detected. By 1974, over 160 separate x-ray objects had been discovered.

In 1971, astronomers began to focus particular attention on a strong x-ray source in the constellation of Cygnus the Swan, shown in Figure 8-1. This source had all the appearances of an invisible starlike object in a binary system. By 1972, it was established that Cygnus X-1 was orbiting a 9th magnitude visible star called HDE 226868. Careful examination of HDE 226868, a single-line spectroscopic binary, revealed that its unseen x-ray–emitting companion was at least six times more massive than the sun. This is exactly what the astrophysicists had been hoping for. Cygnus X-1 was the first bona fide black-hole candidate.

The excitement aroused by Cygnus X-1 was due to earlier calculations by Drs. N. Shakura and R. Sunyaev in Moscow and Drs. J. E. Pringle and M. J. Rees in England. In 1971, before it was known that x-ray sources exist in binary stars, these two teams of astrophysicists performed some detailed calculations concerning the consequences of gas from an ordinary star falling on a

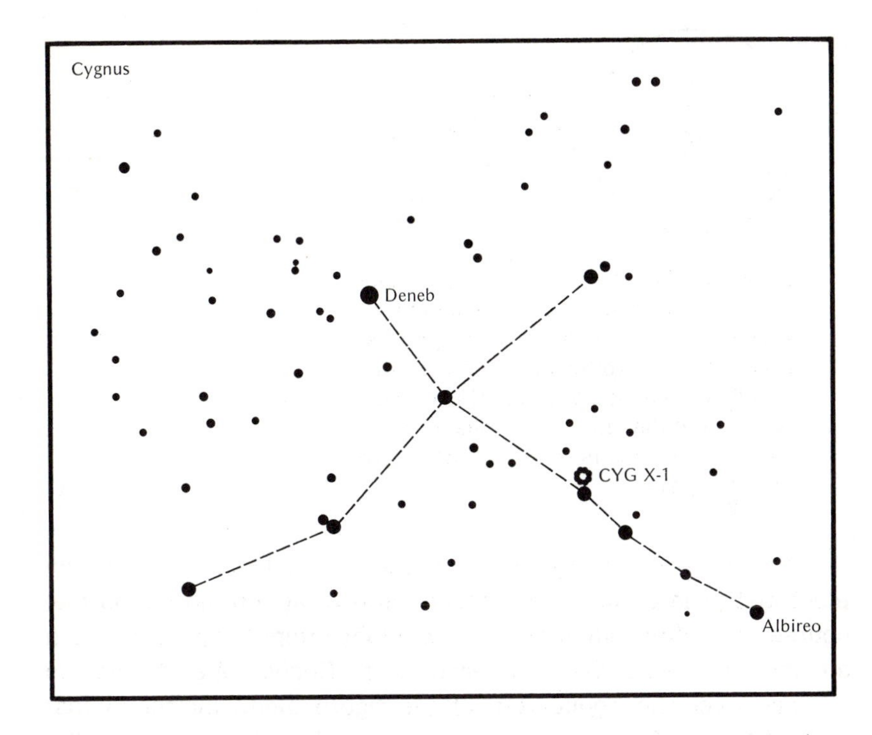

Figure 8-1. Cygnus X-1. *This object* (also known as 3U 1956+35) is an intense source of x-rays. It is believed that the x-rays are produced by gases spiraling into a black hole.

black hole in a binary system. In order to appreciate these theoretical discoveries, we must first review some basic ideas about the dynamics of binary stars.

Ever since the nineteenth century, it was realized that *mass exchange* could occur between stars in a binary system. This phenomenon is due to the fact that the effective gravitational force of a star in a binary system extends over a limited range. Specifically, it is possible to draw a figure eight, which denotes the effective limits of the stars' gravitational field, around the two stars in a binary, as shown in Figure 8-2. Any gas floating around inside one lobe of the figure eight is gravitationally bound to the star in that lobe. Gas floating outside of the figure eight is free to drift off into interstellar space. Each half of the figure eight is called a *Roche lobe* in honor of the nineteenth-century physicist who first discovered their importance.

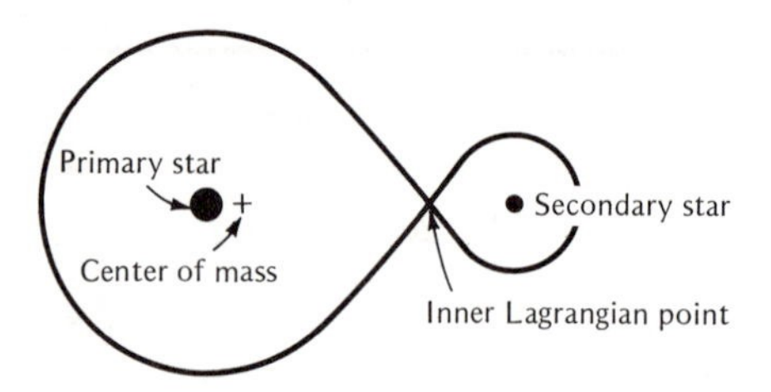

Figure 8-2. The Roche lobes. A figure eight drawn around the two stars in a binary system defines the limits of the gravitational influence of each of the two stars. Gas floating inside a lobe of this curve is bound to the star in that lobe. Gas outside the lobes is free to escape from the binary star.

There are two ways in which a star in a binary system can eject matter into interstellar space. First of all, during the normal course of stellar evolution, a star might simply expand to such an enormous size that it *overflows* its Roche lobe. As gas is dumped over the boundaries of the figure eight, matter is lost from the star. A second method of mass loss involves a *stellar wind.* Astronomers believe that all stars are constantly ejecting high-speed particles (protons and electrons) into space. For example, ever since the early days of the space program, satellites have detected high-speed particles ejected from the sun's surface producing the so-called *solar wind*. Such particles travel so rapidly that they easily pass across the Roche lobes of a binary star. This accounts for a much more gradual rate of mass loss from a star. These two methods of mass loss from a binary system are contrasted in Figure 8-3.

To understand the significance of theoretical work on x-ray binaries during the early 1970s, suppose that one star in a binary system is a black hole. As long as the ordinary star in such a system is not ejecting matter over its Roche lobe, the black hole will be undetectable. But if the normal star is blowing off matter into space, then some of this gas will flow through the crossover point of the figure eight (called the *inner Lagrangian point)* and be captured by the black hole. From the initial work of Shakura, Sunyaev, Pringle, and Rees, it was realized that gases captured by the black hole will go into orbit around the hole, thereby pro-

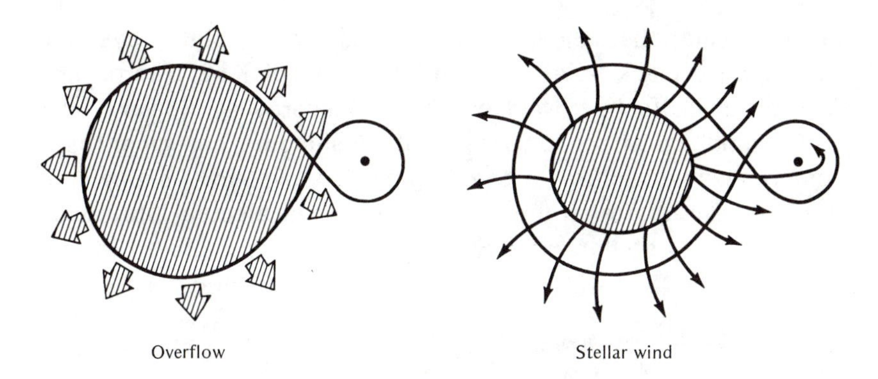

Figure 8-3. Mass loss from a binary star. A star in a binary system can lose matter in one of two ways. By expanding to a very large size, the star can simply overflow its Roche lobe. A more gradual rate of mass loss results from a stellar wind of high-speed particles ejected from the star's surface.

ducing a huge disc. This so-called *accretion disc* is constantly fed by matter flowing from the normal star and orbits the black hole like a giant version of the rings around Saturn.

It is well known that Mercury orbits the sun faster than Pluto. Orbits close to a source of gravity simply require a higher speed than those further away. Similarly, gas in the inner parts of an accretion disc surrounding a black hole revolve around the hole faster than gas in the more remote parts of the disc. The fact that different parts of the accretion disc orbit the black hole at different speeds means that gases at various distances from the hole rub against each other. The rapidly moving gases near the hole rub against the more slowly moving gases further away. The resulting friction causes the accretion disc to heat up. Far from the black hole, the temperature of the gas in the accretion disc can be quite cool, perhaps only a few thousand degrees. But by the time this gas has spiraled into the inner regions of the disc, friction has heated the gas to temperatures in excess of 1 million degrees Kelvin. Anything with a temperature of 1 to 2 million degrees naturally emits x-rays. In other words, the incredibly hot gases in the innermost part of an accretion disc surrounding a black hole should be a powerful source of x-rays.

Careful observations of Cygnus X-1 during the early 1970s prompted several teams of astrophysicists to perform more detailed calculations concerning mass exchange in binary systems containing a black hole. The resulting work of K. S. Thorne and D. Page at Caltech, I. Novikov and A. Polnarev at Moscow, and C. Cunningham at the University of Washington has given us a comprehensive picture of the Cygnus X-1 system. As depicted in Figure 8-4, matter from the normal star is ejected into space by means of a stellar wind. Some of this gas is captured by the

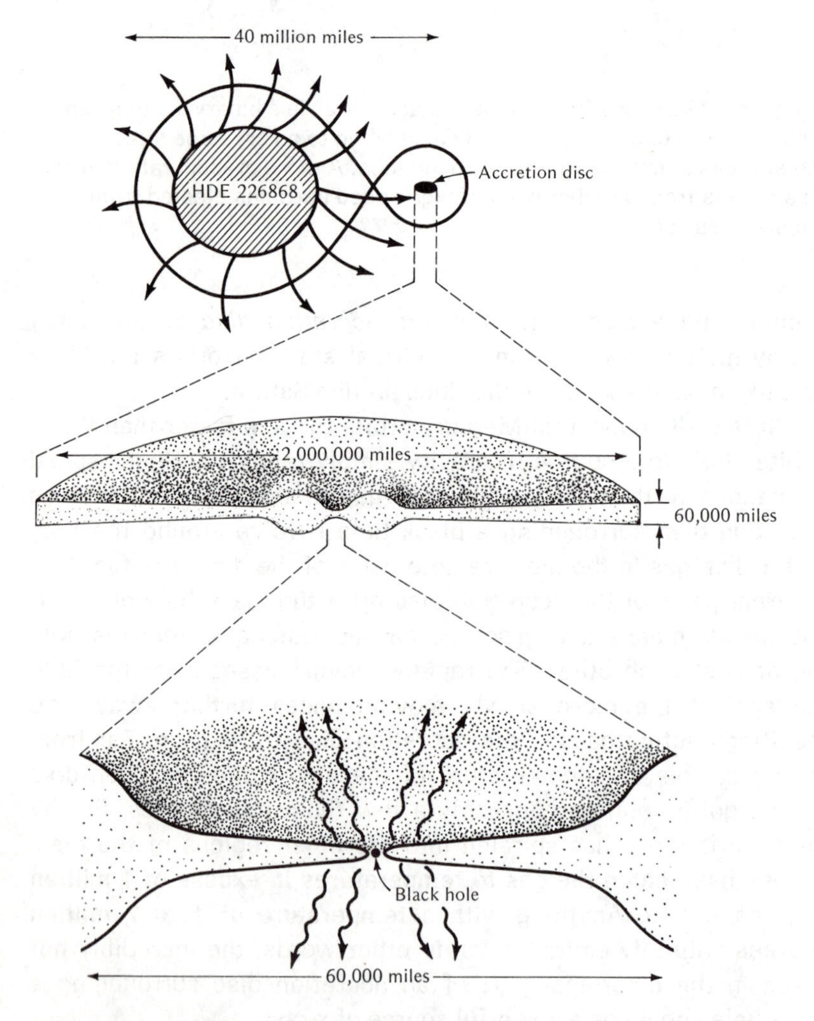

Figure 8-4. The Cygnus X-1 system. Gases ejected from the normal star (HDE 226868) are captured into an accretion disc orbiting a black hole. X-rays are emitted from the hot gases in the innermost parts of the disc.

black hole, producing an accretion disc about 2 million miles in diameter. Friction in the disc causes the gases to heat up as they gradually spiral in towards the black hole. In the innermost parts of the disc, about 200 miles above the hole, the gases are so hot that they naturally emit vast quantities of x-rays. The black hole itself is probably only 20 miles in diameter. Thus the x-rays do *not* come directly from the hole but rather from the hot, inward-spiraling gases, just before they plunge into the hole.

Many astronomers during the mid-1970s have been very cautious in adopting the black-hole interpretation of the x-rays from Cygnus X-1. Taking the position of devil's advocate, they have tried to explain the observations of Cygnus X-1 without resorting to a black-hole interpretation. However, all of these conservative interpretations now look quite artificial and contrived. Indeed, the simplest and most straightforward explanation of Cygnus X-1 is that it is simply a black hole surrounded by an accretion disc. All subsequent observations support this view and, therefore, the discovery of a black hole has gradually gained acceptance.

There are several additional x-ray sources in the sky that have properties very similar to Cygnus X-1. For example, Circinus X-1 (also known as 3U 1516-56) is in a binary system that might contain a black hole. Additional observations with improved spacecrafts will, however, be necessary before firm conclusions can be drawn. Astronomers are anxiously awaiting the launch of HEAO (i.e., the High Energy Astronomy Observatories) scheduled for the late 1970s and early 1980s. These superb spacecrafts will be capable of sensitive observations that may result in the discovery of numerous additional black holes scattered through space.

Galaxies and Quasars

Several hundred years ago it became very evident that there was a great deal more to see in the sky than just the stars and planets. By the turn of the nineteenth century, the quality of astronomical telescopes had improved to such a degree that many astronomers were making extensive observations and catalogs of "fuzzy" objects called *nebulae.* Several of these fuzzy objects turned out to be comets and in one case a new planet. But the great majority of the nebulae remained stationary with respect to the fixed stars, indicating that these nebulae were at extremely great distances from our solar system.

The question of the acutal distances to the nebulae was destined to remain a mystery until the early part of the twentieth century. At that time, the use of photographic techniques along with the new 100-inch telescope at Mt. Wilson permitted astronomers to obtain views of these objects far superior to anything previously available. It was then apparent that all nebulae fell into one of two categories. First of all, there were, within our own galaxy, objects such as the Crab Nebula, the Ring Nebula, and the Orion Nebula that consisted of clouds of gas that were glowing for one reason or another. Today we realize that these objects in this first category include H II regions and nova remnants, the birthplaces and graveyards of ordinary stars. The objects in the second category are somewhat more confusing. They include M31 (the Andromeda Nebula), M87, and M101 and frequently had a spiral or pinwheel appearance.

Although there was little controversy about the approximate locations of the nebulae in the first category, a great debate soon developed concerning the distances to the pinwheel objects. Some people felt that they were within the confines of our own galaxy, while others argued that they were at enormous distances from our own galaxy. Indeed, perhaps they were separate galaxies or "island universes," as suggested many years earlier by Immanuel Kant. Astronomers were therefore faced with the problem of developing new methods of determining distances.

If someone takes a 200-watt light bulb and stands an unknown distance away from you in the night, you can relatively easily determine this unknown distance. If you observe how bright your friend's 200-watt light bulb appears to your eye, and compare this apparent brightness with the real, or absolute, brightness of a 200-watt light bulb, then you readily conclude that there is one and only one possible distance between you and your friend. In other words, by comparing the apparent brightness and absolute brightness of the light bulb, the distance to the light bulb is determined. If, then, astronomers could find some stars for which, for one reason or another, they could determine the real, or absolute, brightness by observing their apparent brightness in the sky, they could easily calculate their distances.

As luck would have it, shortly after the turn of the present century, it was discovered that Cepheid variables possess a unique relationship between their periods and their absolute magnitudes. Cepheid variables are stars that alternately get brighter and dimmer in a very specific fashion. They brighten rapidly and dim slowly within a regular period, which, depending on the particular star, can lie anywhere between about one day and one month. By studying the Cepheid variables in the Magellanic Clouds, it was discovered that the rapidly varying Cepheids have a low real brightness, while the slowly varying Cepheids have a much higher real brightness. This unique relationship between the periods and absolute magnitudes of Cepheid variables, known as the *period-luminosity* law, is exactly what we had been looking for. Astronomers then began searching for Cepheids in the spiral nebulae. Cepheids are easily recognizable because of their "light curve," or the characteristic way in which they vary their brightness. After some observations with a telescope and a clock, the period of a Cepheid is easily measured. Then, by referring to the period-luminosity law, the real brightness of the star in question can be deduced. By comparing this absolute magnitude with the apparent brightness of the star, the distance is readily determined. By performing these and similar observations, it was soon concluded that the spiral nebulae were indeed at enormous distances from us, far beyond our own galaxy.

It was therefore obvious that spiral nebulae, or *galaxies* as they had come to be called, were an entirely new class of astronomical object. Astonomers were seeing much further out into the universe than had previously been supposed, and many scientists diligently turned their attention to these objects. In particular, Dr. Edwin Hubble discovered that all galaxies could be classified into one of four categories: ellipticals, spirals, barred spirals, and irregulars. Ellipticals have the appearance of amorphous blobs with no visible structure. They can be circular in appearance or quite flattened. Spiral galaxies show the familiar spiral arms originating at the nucleus of the galaxy and winding outward. Barred spirals are similar to spirals except that the spiral arms originate at the ends of a bar running through the center of the galaxy rather than at the nucleus. And finally, anything that does not fit into the previous three categories usually has a very odd or disturbed appearance and is called an irregular galaxy.

By examining the spectra of galaxies, Dr. Hubble also discovered that the nearby galaxies are moving away from us slowly, while the more distant galaxies are rushing away from us much more rapidly. This so-called Hubble law, of which we have spoken earlier, was the first indication that the universe as a whole is expanding. In addition, this Hubble law provided astronomers with another important distance indicator. For example, if you wanted to find the distance to a particular galaxy, you would measure the redshift of the spectral lines in the spectrum. This redshift would correspond to a certain recessional velocity which, by the Hubble law, would in turn correspond to a particular distance. In this way, astronomers had become accustomed to interpreting the redshifts of galaxies as a cosmological effect reflecting the expansion of the universe. This is roughly where our knowledge stood around 1960. Everything seemed to make sense. And then quasars came along!

During the 1930s, it was discovered that radio waves from space were being detected here on earth. The implications of this discovery were left untouched for almost a decade due to the intervention of World War II. In the late 1940s and early 1950s, however, astronomers found that many of the wartime advances in electronics and electrical engineering could be directly applied to the task of building radio telescopes. For the first time, scientists had the opportunity to see what the universe looks like at wavelengths far removed from ordinary light. And indeed the view is very different. Radio astronomers promptly began making catalogs that recorded the positions and strengths of the sources they discovered. The most famous work produced as a result of these efforts was the *Third Cambridge Catalogue.* (All entries in this catalog begin with the symbol 3C, just as the entries in Messier's catalog begin with M and those in the *New General Catalogue* begin with NGC.)

Given the positions of radio sources, such as from the 3C catalog, optical astronomers turned their telescopes to the relevant portions of the sky in the hope of discovering some unusual visible objects. In many cases, their efforts were quite successful, and most radio sources came to be identified with diffuse gaseous objects supposed to contain extensive magnetic fields. This made sense. As the charged particles in an ionized gas or plasma move about an extensive magnetic field, the particles (electrons) will radiate radio waves due to *synchrotron emission.* In addition, the regions emitting radio waves had to be very large in order to produce enough energy to be detected here on earth. Indeed, up until 1960 all optically identified radio sources were associated with large diffuse gaseous objects ranging from supernova remnants to entire galaxies.

In 1960, however, radio astronomers turned their attention to a few of the sources in the *Cambridge Catalogue*, such as 3C 48 and 3C 273, which did not seem to be extended over a large region of the sky. In particular, 3C 273 is located in Virgo near the ecliptic and is occasionally occulted by the moon. Such an occultation occurred in 1960, and, to everyone's amazement, when the moon passed over the position for 3C 273, the radio emmissions ceased abruptly, after producing interference bands. This was conclusive proof that 3C 273 is a point source of radio noise! From knowing the precise position of the moon during the occultation, it was fairly easy to pin down a precise location for 3C 273. At this location was a 12.6 magnitude bluish "star." Even more surprisingly, when optical spectra were obtained for 3C 273, no one could identify the spectral lines! Astronomers simply had never seen a spectrum like this before. By 1963 the situation was four times as bad because three more of these bluish starlike objects emitting radio noise had been discovered. In all cases, the spectra remained a complete mystery. By this time, these sources had been christened *quasi-stellar objects,* or *quasars* for short.

The first real advance in understanding quasars was made by a young Dutch astronomer, M. Schmidt, at Caltech. He discovered that if he assumed quasars had truly enormous redshifts, then he could in fact explain the spectral lines as having been shifted all the way from the blue and even ultraviolet parts of the spectrum. The reason astronomers had been puzzled by quasar spectra is simply that they had never before seen such huge redshifts. Today it is quite common for astronomers to be observing quasars in which the redshift is so great that the spectral lines they see have come all the way from the invisible ultraviolet. If we interpret the redshifts in terms of a velocity, quasars must be moving away from us at up to 90 percent of the speed of light. Furthermore, by applying the Hubble law, these velocities imply distances of billions of light years, far beyond any known galaxies.

Ever since the early 1960s, the entire astronomical community has been reeling under the onslaught of one mind-boggling observation after another. In fact, it would seem that by any reasonable standards whatsoever, quasars should simply not exist. First of all, astronomical radio sources should be large diffuse objects. The theoretician is at a complete loss to explain how a compact stellarlike object can emit enough radio noise for that noise to be detected on earth. Indeed, during the mid-1970s evidence accumulated that suggests that the radio-emitting regions in a quasar are only a few billion miles across—comparable to the size of our solar system. Furthermore, if we interpret the redshifts as cosmological, then quasars should be so far away that it should be totally impossible to detect them, either with radio telescopes, the 200-inch telescope, or anything else. Yet there they are, and by 1970 the list of known quasars ran into the hundreds and continues constantly to expand.

Virtually all attempts at explaining quasars have met with dismal failure. If you assume that the redshift is cosmological, then quasars are so far away that unbelievably violent sources of energy totally unknown to modern scientist must be postulated to explain why quasars shine so brightly. If we assume that the redshift is due to general relativity in the form of a gravitational redshift, then quasars should rapidly form black holes and disappear from the universe. If we assume that quasars are really very nearby objects that happen to be moving at very high speeds, then we cannot explain why we do not see any comparable blueshifts, because, after all, there ought to be some quasars coming toward us.

In the late 1960s, an interesting trend in the observations began to emerge. Certain observations suggested that quasars seemed to be found near galaxies, especially near peculiar or exploding galaxies. For example, there are about 50 quasars in the *Third Cambridge Catalogue.* Surprisingly, about half a dozen of these lie extremely close to galaxies listed in the *New General Catalogue* (for example, NGC 4651 and 3C 275.1, NGC 3067 and 3C 232, NGC 4138 and 3C 268.4, NGC 5832 and 3C 3091, and NGC 7413 and 3C 455). The probability for such a close association to occur at random between numbers of the NGC and 3C catalogs is about 1 in 10,000. This cannot easily be ignored.

Meanwhile, many astronomers began turning their attention to a variety of peculiar galaxies. For example, in the 1940s C. Seyfert had discovered a number of galaxies that appeared to have extremely bright nuclei but very poorly developed dim spiral arms. These Seyfert galaxies, as well as the N-type and compact galaxies of Zwicky, were discovered to have many properties halfway between quasars and ordinary galaxies. For example, the redshifts of Seyferts, N-types, and compacts lie roughly between the high redshifts for quasars and the more normal redshifts for ordinary galaxies. Are we seeing an evolutionary effect? Do quasars become

Seyferts, which become ordinary galaxies? The idea that quasars are ejected out of the nuclei of active, or exploding, galaxies was largely ignored by most astronomers because they failed to observe any blueshifted quasars. After all, we are not in any special place in the universe, and some of these quasars ought to be coming toward us. But even more impressive data continued to pour in.

Ignoring the question of quasars for a moment, it has become extremely apparent that active galaxies do in fact explode and eject great quantities of matter. There are first of all many impressive photographs showing exploding galaxies, such as M82, NGC 1275, and M87. In the case of M82, not only do we see a very distorted object, but the nearby galaxy M81 has obviously been affected by its exploding neighbor. Close examination of M81 reveals a series of ripples that possibly are the result of shock waves from the exploding M82. In the case of NGC 1275, R. Lynds has produced a remarkable series of photographs in the light of the hydrogen atom. These photographs show an object that looks more like the Crab Nebula than a galaxy. M87 in the Virgo cluster looks like an ordinary giant elliptical galaxy until you take a short-exposure photograph. Such a photograph shows only the bright central part of the galaxy, along with a huge jet of gas surging out of the nucleus. Improved photographs by Halton Arp, taken using the 200-inch telescope at Palomar, show this jet to consist of blobs or condensations of matter, as well as of counterjet pointing in the opposite direction.

Not to be outdone by the optical astronomers, radio astronomers in the mid-1960s began reporting great quantities of data strongly suggesting violent cosmic events. For example, one of the strongest radio sources in the sky, Cygnus A (also known as 3C 405), shows the characteristic dumbbell shape of the radio objects associated with cataclysmic events. A radio map of Cygnus A is given in Figure 9-1. Radio observations of this type are thought to be the evidence for the simultaneous ejection of two large masses in opposite directions out of an exploding galaxy. Calculations indicate that the masses of the ejected blobs are probably in the order of 100 million suns and are filled with very high-speed (relativistic) electrons that radiate the radio noise we observe.

Frequently, instead of a simple dumbbell shape, which would

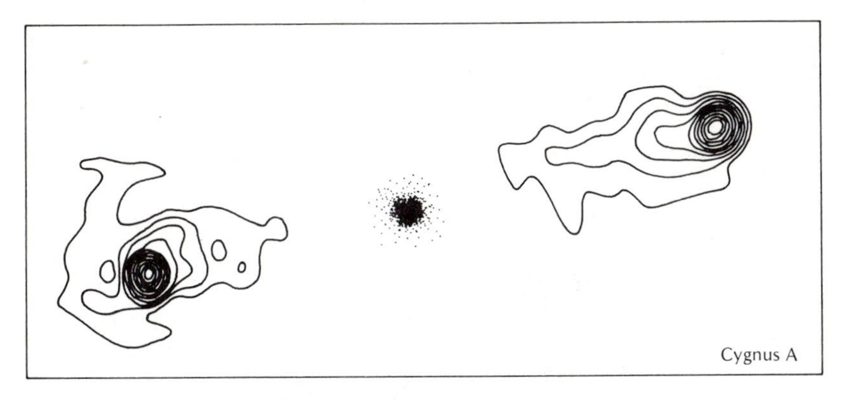

Figure 9-1. A radio map of Cygnus A. Cygnus A (also known as 3C 405) is one of the brightest radio sources in the northern sky. Radio astronomers present their data by drawing lines of constant signal intensity. The resulting charts look like a surveyor's contour map. Careful observation of this object with large radio telescopes reveals that this source is double. Most of the radio noise comes from two blobs on either side of a peculiar galaxy that is sketched in.

indicate a single explosion, there are much more complex shapes, suggesting two or more explosions. In fact, about 10 percent of the dumbbell sources are actually quadruple. A good example of this phenomenon is the famous object, Centaurus A, seen in the southern sky. At first glance, Cen A has the appearance of an ordinary dumbbell source. With the aid of optical telescopes, Cen A was conclusively identified with NGC 5128, which looks like an elliptical galaxy with a spectacular dust lane running through its center. Closer examination with a radio telescope reveals two more blobs inside the region of the optical image that seem to be moving outward in roughly the same direction as the original two blobs, as shown in Figure 9-2. This direction is very nearly perpendicular to the dust lane seen in photographs of NGC 5128.

More recently, several double and multiple radio sources that dwarf our own galaxy have been discovered. As shown in Figure 9-3, 3C 236 and DA 240 are hundreds of times larger than our entire galaxy, which is only 100,000 light years in diameter.

Moving into the 1970s, astronomers gradually became more willing to accept the idea that active galaxies can eject vast quantities of matter in a violent fashion. But the idea that quasars,

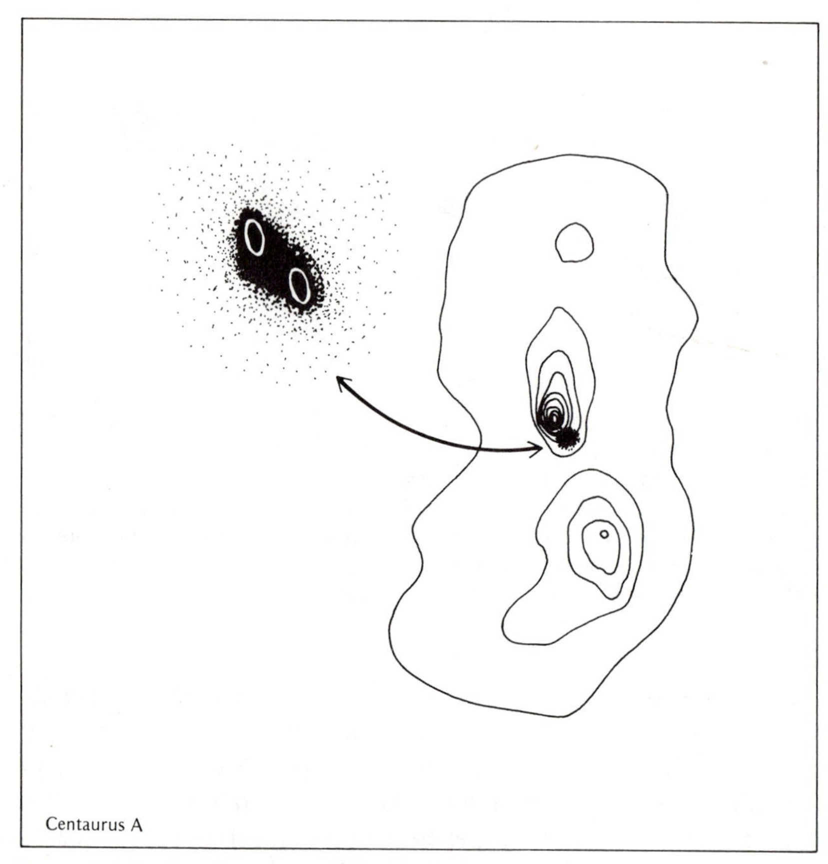

Figure 9-2. Two radio maps of Centaurus A. Centaurus A is a very bright radio source in the southern sky and is associated with the exploding galaxy NGC 5128. On the right we see an overall map of the entire source. Most of the radio noise comes from two blobs located on either side of the galaxy. Close examination of the central region (shown on the left) reveals two more blobs located inside NGC 5128. We, therefore, refer to Cen A as a quadruple source.

with their enormous redshfits, were ejected in a similar manner remained repugnant for the reasons previously outlined. Nevertheless, impressive evidence continued to mount, due primarily to the persistent efforts of Halton Arp at the Hale Observatories. Specifically, Dr. Arp has pointed out numerous alignments of exploding galaxies and quasars that could not possibly be due to chance alone. For example, extending southwest from the exploding galaxy NGC 520, there are four quasars that lie in almost a perfectly straight line! Maps of this region of the sky are given in Figure 9-4.

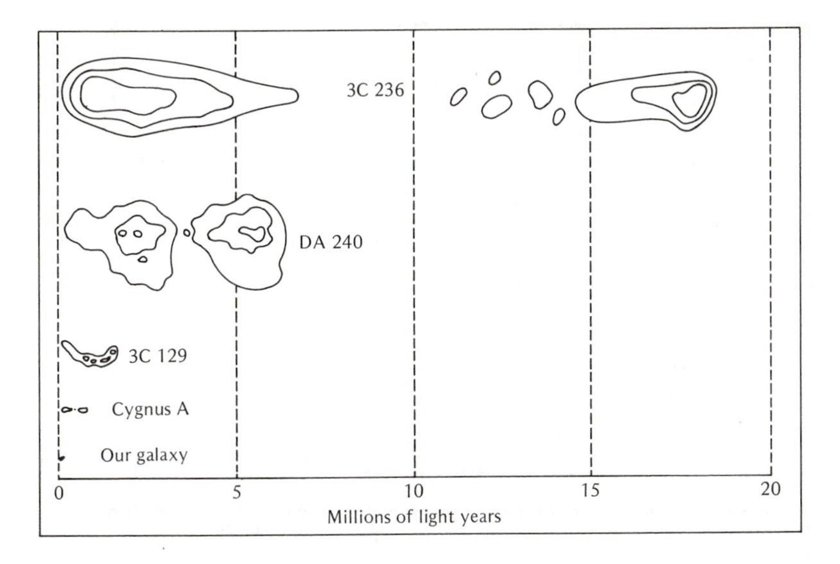

Figure 9-3. Giant radio galaxies. Four radio galaxies are compared with our own galaxy. Most double radio galaxies are about the same size as Cygnus A, about 600,000 light years across. (Adapted from Drs. R. G. Strom, G. K. Miley, and J. Oort, *Scientific American,* Vol. 233, No. 2, August, 1975).

By far the most impressive evidence linking quasars and exploding galaxies came in 1971, when Dr. Arp succeeded, with the 200-inch telescope, in obtaining a photograph that showed a bridge of gas connecting the peculiar galaxy NGC 4319 and the quasar known as Markarian 205. The redshift in the spectrum of NGC 4319 indicates a recession velocity of 1050 miles per second; the redshift of the quasar Markarian 205 corresponds to a recessional velocity of 13,000 miles per second. In view of fact that these two objects appear side by side in the photograph and are connected by a bridge of gas, the Hubble law cannot apply to both the quasar and the galaxy at the same time.

Clearly, there is something very wrong. The Hubble law does not seem to apply to quasars, and all other familiar physical processes usually employed to explain redshifts either break down completely or simply ridiculous or extremely artificial situations. These difficulties were compounded even more in 1971 by the work of the Finnish astronomer T. Jaakkola. Dr. Jaakkola took a good

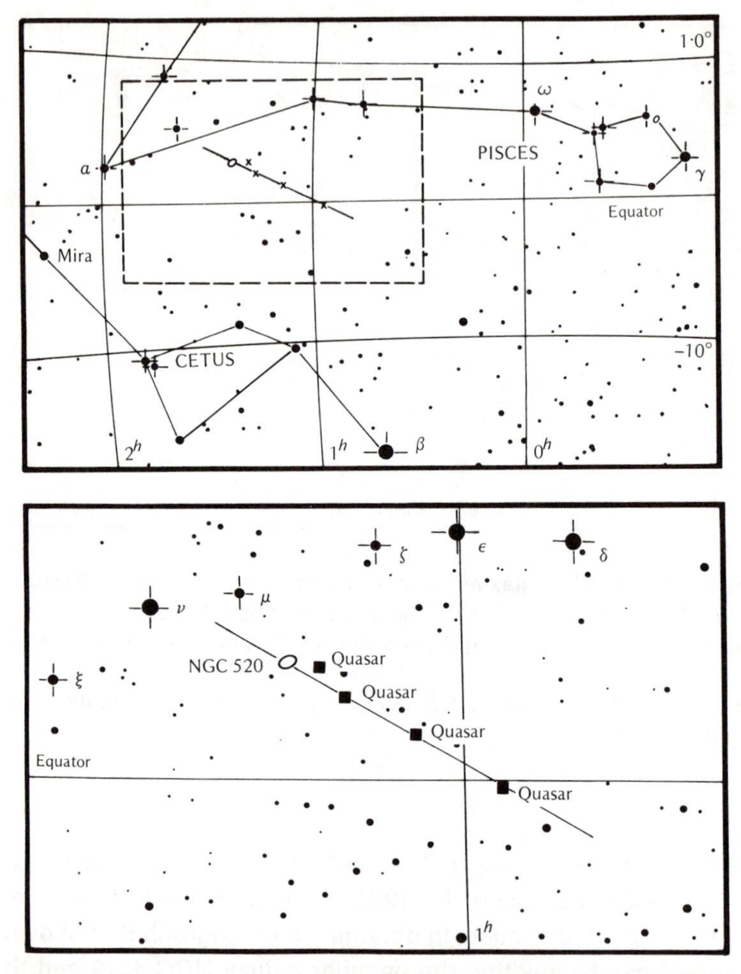

Figure 9-4. Maps of the sky near NGC 520. The upper chart shows a large region of the sky including portions of Pisces and Cetus. The lower chart is a more detailed map of the sky around the exploding galaxy NGC 520. As shown on this chart, there are four quasars that lie very nearly on a straight line to the southwest of this galaxy.

hard look at well-known clusters of ordinary galaxies. Much to everyone's surprise, he found that ordinary spiral galaxies with well-developed spiral arms have, on the average, slightly higher redshifts than ellipticals or spirals with very tightly wound spiral arms in the same cluster! This phenomenon of discrepant redshifts becomes even more pronounced if we include peculiar galaxies.

For example, VV 172 (the 172nd object listed in a catalog prepared by the Soviet astronomer Vorontsov-Velyaminov) shows five galaxies in almost perfect alignment. One of the five galaxies has a much higher redshift than any of the remaining four. In addition, the famous Stephan's Quintet (NGC) numbers 7319, 7318A, 7318B, 7317, 7320) shows a similar disparity in redshift. One member, NGC 7320, has a redshift that is only one-eighth as large as the redshifts of any of the remaining four galaxies.

In 1973, Dr. Arp tackled the problem of Stephan's Quintet head-on. Many galaxies contain H II regions. An H II region is a large cloud of glowing hydrogen gas such as the famous Orion Nebula. Assuming that all H II regions have roughly the same diameter, it should be possible to deduce the distance to a galaxy by observing the apparent sizes of H II regions in that galaxy. H II regions in nearby galaxies should appear large through a telescope, while H II regions in a distant galaxy should look very small.

If all the galaxies in Stephan's Quintet are at the distances indicated by their redshifts, then NGC 7319, 7318A, 7318B, and 7317 should be eight times farther away from us then NGC 7320. Consequently, the H II regions in NGC 7320 should appear eight times larger than the H II regions in any of the four high-redshift galaxies.

Using a series of filters and the 200-inch telescope at Palomar, Dr. Arp carefully photographed the H II regions in Stephan's Quintet. Figure 9-5 shows a map of Stephan's Quintet (compare with Plate 13) and the locations of all the H II regions that Arp found. The open circles indicate the sizes and locations of the H II regions in the low-redshift galaxy (NGC 7320), while the filled circles indicate the sizes and locations of the H II regions in the high-redshift galaxies (NGC 7318 A and B). The striking and controversial result of Arp's observations is that all the H II regions seem to have roughly the same apparent size. The traditional interpretation of the redshifts in Stephan's Quintet would have predicted that the open circles in Figure 9-5 should —on the average—be eight times larger than the filled circles. Glancing at Arp's results, we see that the distribution in the sizes of H II regions for both the low- and high-redshift galaxies are essentially the same. Dr. Arp therefore concluded that *all* five galaxies are actually at the distance of NGC 7320 and that the remaining four members of Stephan's Quintet possess anomalous, excess redshifts.

If Dr. Arp's intriguing observations are confirmed by future investigations, then there must be something more to the redshifts of galaxies than is given by the Hubble law. Indeed, it may be necessary to search for new laws of physics to explain anomalous high redshifts observed in galaxies and quasars.

In the early 1970s, F. Hoyle and J. V. Narlikar of the Institute of Theoretical Astronomy at Cambridge University succeeded in developing a new and complete reformulation of general relativity that may be extremely relevant to explaining anomalous redshifts. This theory takes the position that the very masses of atoms have their origin in the farthest reaches of the universe. In other words, a proton, for example, has a certain mass because the matter in the rest of the universe has a certain distribution. As the matter in the universe changes, presumably the masses of atomic particles would also change. This theory, if correct, would have tremendous implications for all of physical science.

The Hoyle-Narlikar theory seems to hold great promise for the enigma of the quasars and related problems with redshifts in galaxies. Suppose that matter is being created in the nuclei of exploding galaxies. As the newly created matter emerges from galactic nuclei, it can "see" only a relatively small fraction of the rest of the universe. The matter in the farthest reaches of the universe has not yet had the opportunity to communicate its existence to the newborn atoms. These atoms, therefore, have smaller masses than the much more ancient atoms with which we are familiar. If the electrons in an atom have a lower mass than usual, the electron orbits will be more widely spaced than usual, and all the spectral lines arising from electron transitions will be greatly redshifted. As these new atoms become older, their masses gradually increase because more and more of the outside

Figure 9-5. H II regions in Stephan's quintet. The upper diagram is a map of Stephan's Quintet (compare with Plate 11). The lower diagram, drawn on the same scale as the map, shows the locations and sizes of the H II regions. Open circles denote the H II regions in the low-redshift galaxy (NGC 7320), while the filled circles denote the H II regions in the high-redshift galaxies (NGC 7318 A and B). (Adapted from Dr. H. C. Arp)

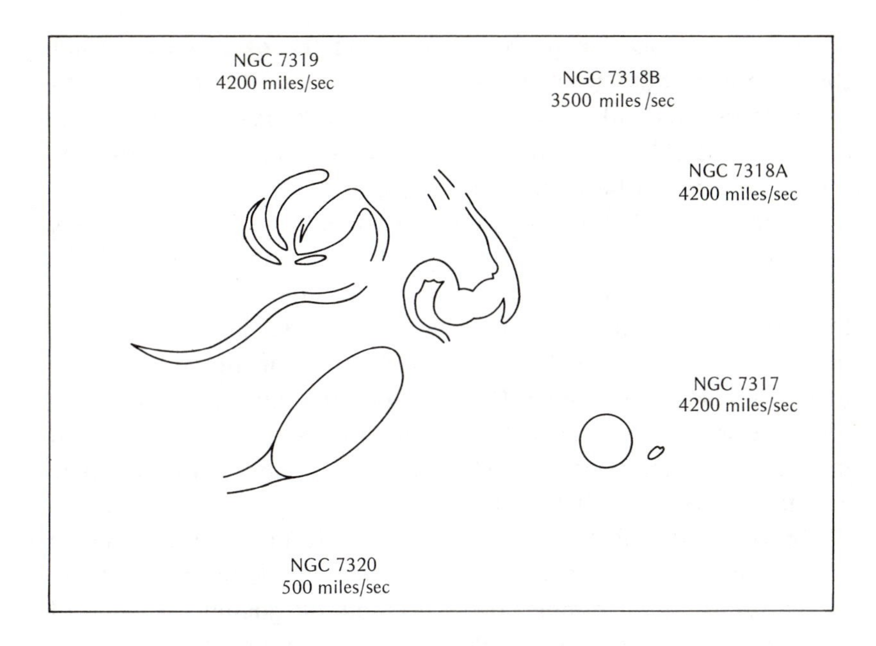
NGC 7319
4200 miles/sec
NGC 7318B
3500 miles /sec
NGC 7318A
4200 miles/sec
NGC 7317
4200 miles/sec
NGC 7320
500 miles/sec

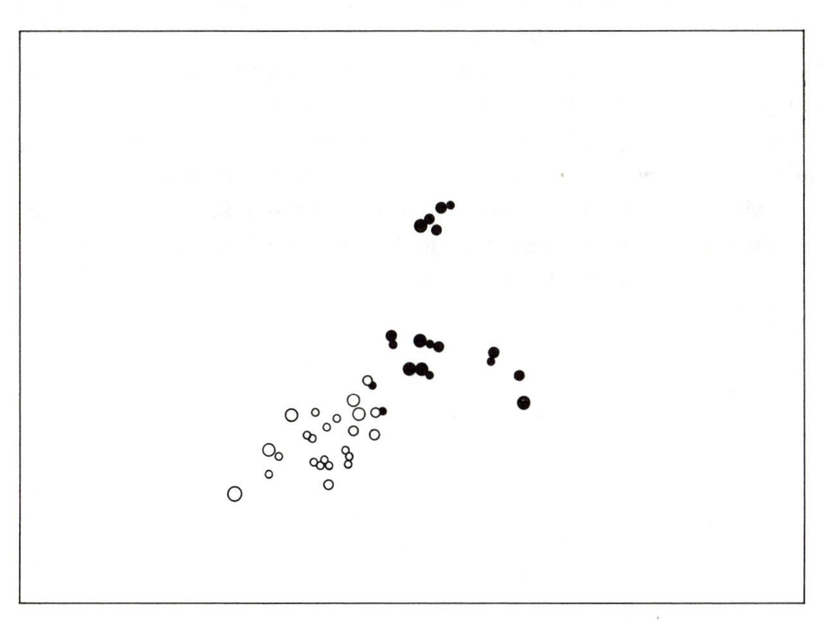

universe has made its existence known to the new matter. When the matter ejected from active galaxies ages, its redshift gradually decreases as the masses of the particles asymptotically approach the values found in our experiments in laboratories with the ancient matter here on earth.

Consider matter gushing up into our universe out of a white hole at the center of an exploding galaxy. In many respects, the emergence of material into this region of space is analogous to the creation of matter. As this ejected material passes the final, outer event horizon, for the first time the rest of the universe can begin to make its presence known. The atomic particles of which this matter is composed will have low masses, which will result in extremely large redshifts. With age, the redshifts will decrease as the masses gradually increase.

Perhaps, therefore, the anomalous redshifts that have been so puzzling to astronomers are really trying to tell us the very history of matter. Matter, either newly created or gushing up out of wormholes from some other universe, first emerges as compact quasars with high redshfits. The quasars age and begin to develop spiral arms. At this point they look like Seyfert, N-type, and compact galaxies with moderately high redshifts. Still much later, with more developed spiral arms they become full-fledged spiral galaxies that ultimately evolve into elliptical galaxies.

Much remains to be done in exploring these fascinating ideas. As a result, it may happen that no concept in physical science will be left untouched by the dramatic revolution in our thinking about the universe.

ELLIPTICAL GALAXY
(M 87 also known as NGC 4486)

SPIRAL GALAXY
(M 74 also known as NGC 628)

BARRED SPIRAL GALAXY
(NGC 1300)

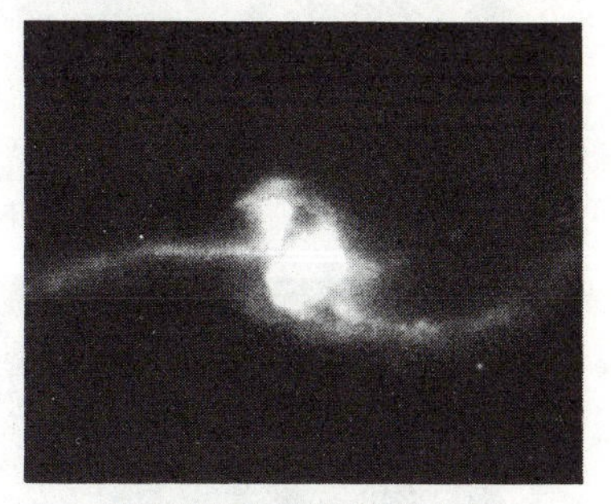

PECULIAR or IRREGULAR
GALAXY (NGC 2623)

Plate 1. Four types of galaxies. There are four typical photographs showing the basic types of galaxies. M87 is an elliptical in the Virgo Cluster. The spiral galaxy M74 is in Pisces. NGC 1300 is a barred spiral in Eridanus, and NGC 2623 is a peculiar galaxy in Cancer. (Hale Observatories Photographs)

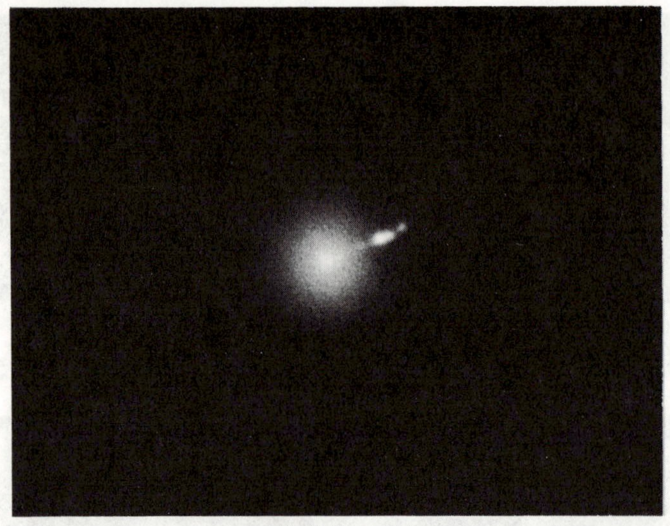

M 87 (NGC 4486)

M 81 (NGC 3031) and M 82 (NGC 3034)

Plate 2.

The jet in M87. A short exposure photograph M87 reveals a huge jet surging up from the nucleus of the galaxy. (Lick Observatory Photograph)

Two galaxies in Ursa Major. M82 (on the right) is an exploding galaxy. The spiral galaxy M81 is nearby. Shock waves from the explosion are now passing through M81. (Hale Observatories Photograph)

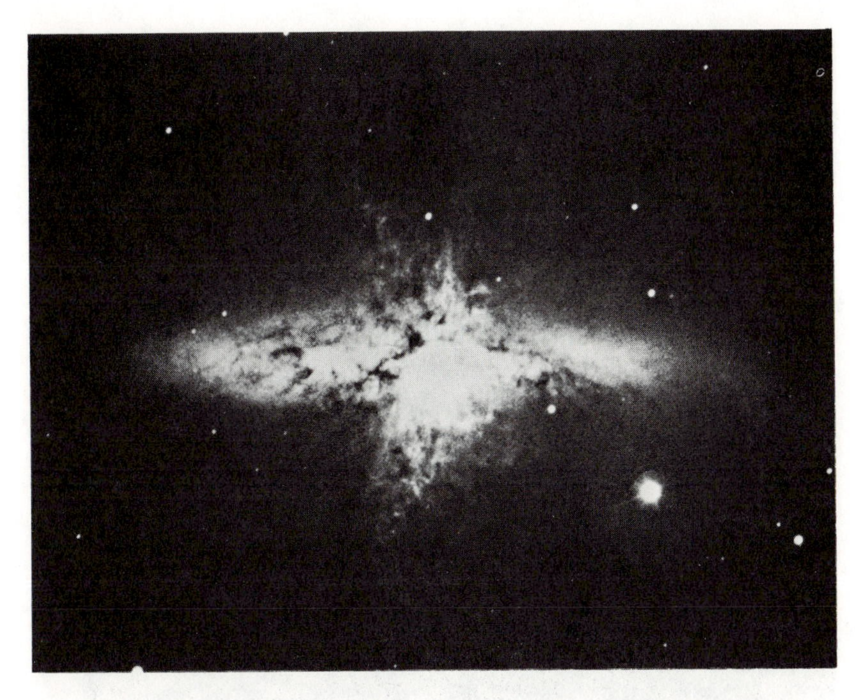

Plate 3. The exploding galaxy M82. This photograph of M82 was taken in the light emitted by hydrogen atoms (H-alpha). Filaments of gas extending outward 10,000 light years from the center of the galaxy can be seen. (Hale Observatories Photograph)

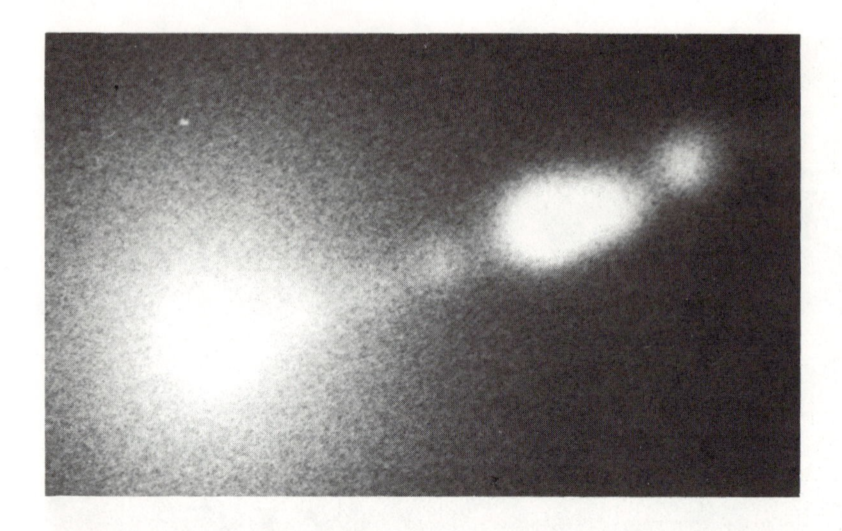

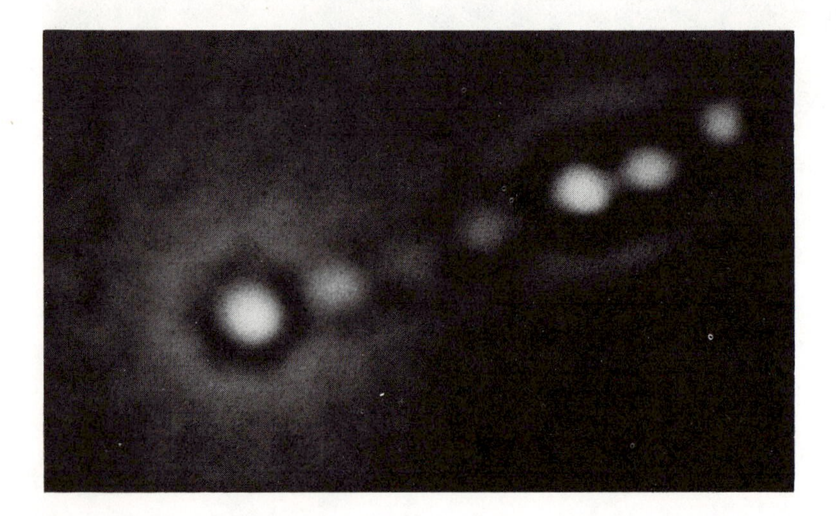

Plate 4. Details of the M87 jet. The upper photograph of the jet in M87 was taken at the 200-inch telescope by H. C. Arp. In 1975, Jean J. Lorre at Jet Propulsion Laboratory processed this photograph through the same computer used to "enhance" the photographs of planets sent back by Mariner spacecrafts. The final result is shown in the lower photograph. Notice the remarkable increase in detail. (Hale Observatories; Jet Propulsion Laboratory)

NGC 1275

Plate 5. The exploding galaxy NGC 1275. This photograph of NGC 1275 in Perseus was taken in the light emitted by hydrogen atoms (H-alpha). This exploding galaxy is a strong source of radio noise and x-rays. (Kitt Peak National Observatory Photograph by Dr. Lynds)

Plate 6. The exploding galaxy NGC 1097. Careful examination of this galaxy by Dr. H. C. Arp in 1976 revealed that an explosion in the galaxy's nucleus had disrupted the spiral arms. The photograph at the top was taken in the light of the hydrogen atom. Notice that one of the spiral arms is broken while the opposite arm is missing a large section. The photograph at the bottom is an overexposed composite of many photographic plates and reveals several jets. The photograph at the bottom shows a much larger portion of the sky than does the photograph at the top. (Cerro Tololo Observatory Photograph by Dr. Arp)

Plate 7. Cygnus A. This is a photograph of one of the strongest sources of radio noise in the northern sky. Radio observations indicate that this object is a dumbell source. (Refer to Figure 9-1 for a radio map of this part of the sky.) (Hale Observatories Photograph)

Plate 8. Centaurus A (NGC 5128). This is a photograph of one of the strongest sources of radio noise in the southern sky. Radio observations indicate that this object is a quadruple source. (Refer to Figure 9-2 for a radio map of this part of the sky.) (Hale Observatories Photograph)

Plate 9. The Seyfert galaxy M77. M77 (also known as NGC 1068) is a Seyfert galaxy in Cetus. The nucleus of the galaxy is very bright and its spectrum shows broad emission lines. This galaxy is also a radio source as well as a strong emitter of infrared radiation. (Lick Observatory Photograph)

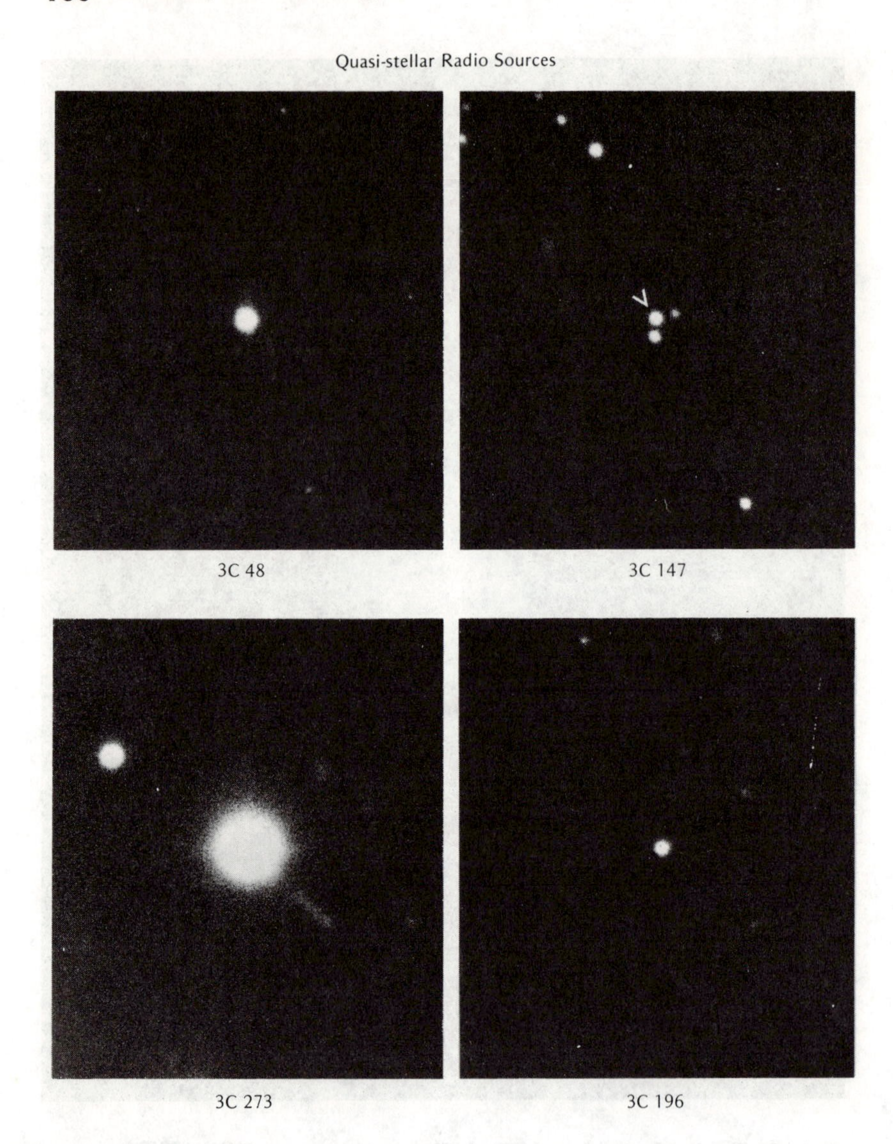

Plate 10. Four quasars. These are photographs of four of the first-known quasars. 3C 48 is in Triangulum (16th magnitude, $z = 0.367$). 3C 147 is in Auriga (17th magnitude, $z = 0.545$). The famous jet can be seen alongside 3C 273 in Virgo (13th magnitude, $z = 0.158$). 3C 196 is in Lynx and has the highest redshift of these four (18th magnitude, $z = 0.871$). (Hale Observatories Photographs)

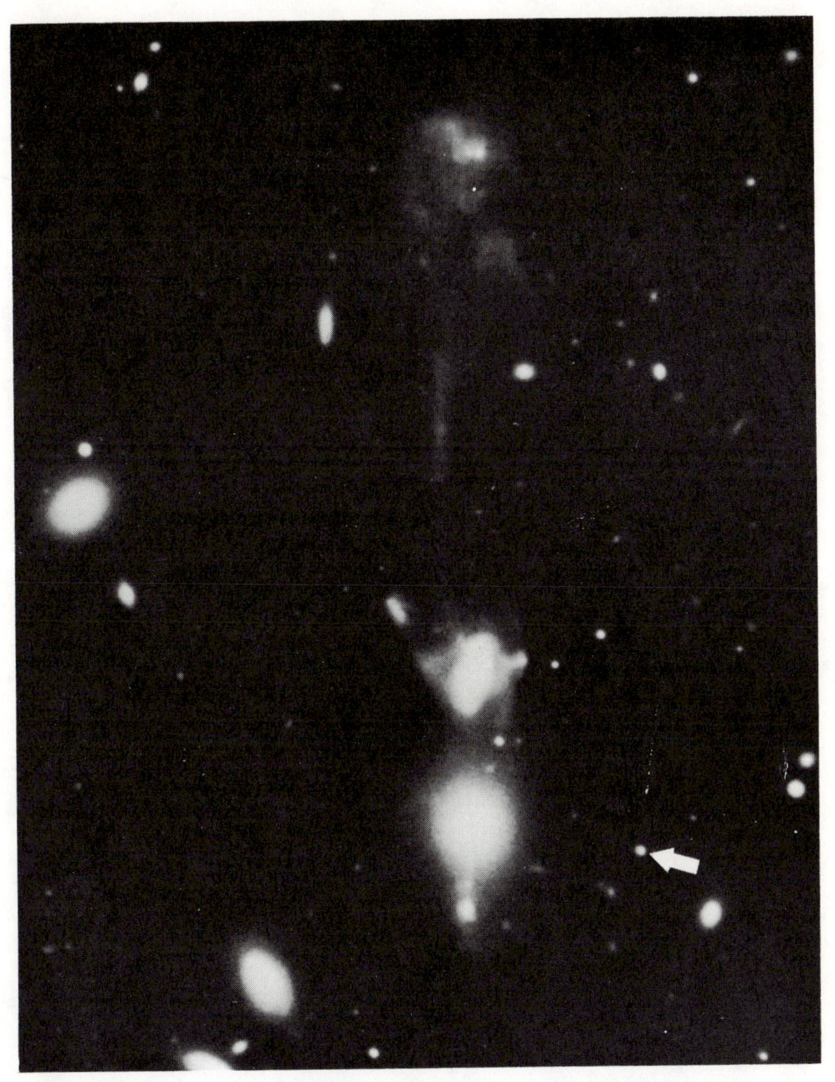

NGC 3561 and quasar

Plate 11. NGC 3561 and a quasar. This photograph of the exploding galaxy NGC 3561 clearly shows the ejection of luminous material. To the west of the spherical galaxy there is a quasar (indicated by arrow) that has an enormous redshift ($z = 2.19$). (Hale Observatories Photograph by Dr. Arp)

Plate 12.

The exploding galaxy NGC 520. Four quasars lie very nearly in a straight line extending to the southwest of this galaxy (see Figure 9-4 for star charts of this part of the sky). (Hale Observatories Photograph by Dr. Arp)

The chain of galaxies VV 172. This chain of five galaxies is located between Draco and Camelopardus. The compact galaxy of this group, one down from the top, has a much higher redshift than any of the other four members. (Hale Observatories Photograph by Dr. Arp)

A bridge between a galaxy and a quasar? These two photographs show an actual physical connection between the galaxy NGC 4319 and the quasar Markarian 205 (indicated by white arrows). The photograph on the left was a four-hour exposure on a baked plate. The photograph on the right is a thirty-minute exposure in H-alpha using an image intensifier. (Hale Observatories Photograph by Dr. Arp)

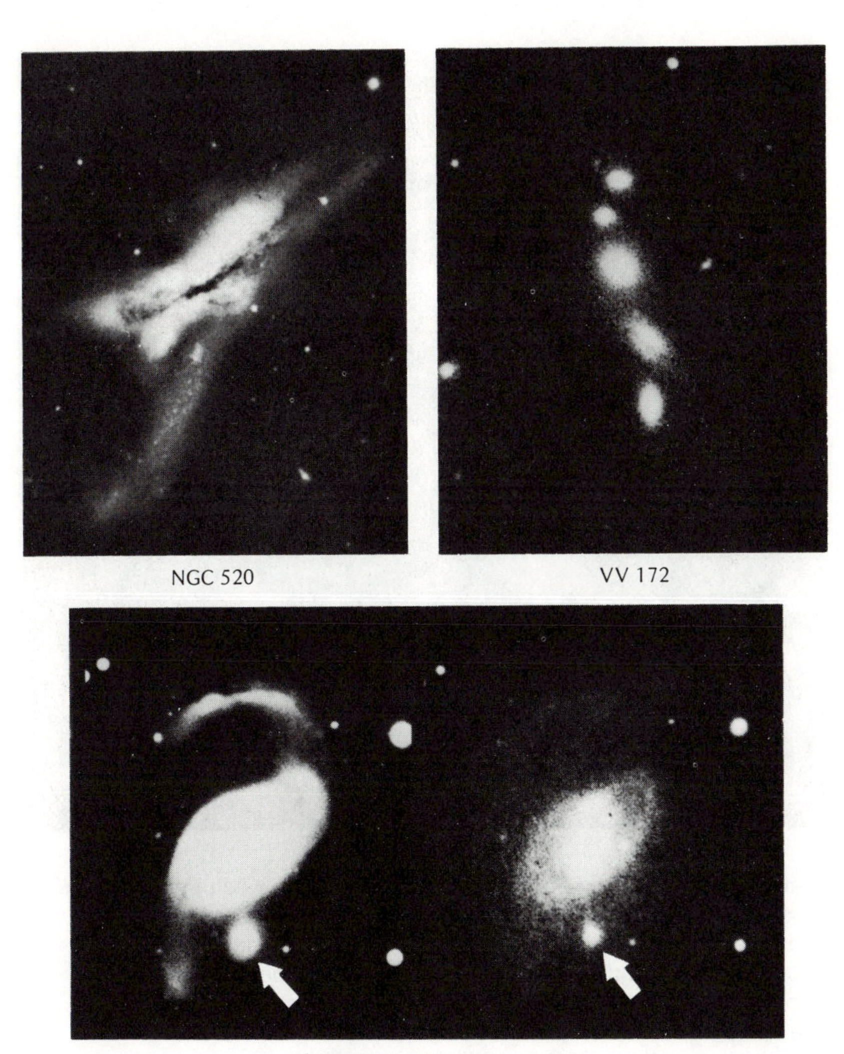

NGC 520

VV 172

NGC 4319 and Markarian 205

Plate 13. Stephan's quintet. The four compact galaxies in this cluster have high redshifts. The fifth galaxy (lower left) has a much lower redshift. Dr. Arp studied the sizes of H II regions in both types of galaxies. Since all the H II regions seem to have roughly the same apparent size, Arp concluded that all the galaxies must be at roughly the same distance from earth. Compare with Figure 9-5. (Hale Observatories Photograph)

10

Gravitational Waves

Einstein's general theory of relativity, which treats the gravitational field as a curvature of four-dimensional space-time, predicts a variety of unusual phenomena. For example, every object that has mass causes the surrounding space-time to become warped. Every time an object moves, this curvature of space-time must readjust to the new configuration of matter. These adjustments of space-time are called *gravitational waves* and move through space at the speed of light. Consequently, every moving object emits gravitational radiation. The earth revolving in its orbit about the sun, a ball bouncing on the floor, a person waving a hand—all such actions emit gravitational waves.

Compared to electromagnetic radiation (light, x-rays, radio waves, and so on, emitted by moving electric charges), gravitational waves are unbelievably weak. One of the reasons for this is that gravitational forces are much weaker than electromagnetic forces. To compare the relative strengths of these two types of forces, consider two electrons. These electrons have mass and charge and exert gravitational and electric forces on each other.

The strength of the gravitational force between two electrons separated by one one-hundredth of an inch is equal to the strength of the electric force between the same two electrons separated by 50 light years (300 trillion miles)! As a result, gravitational waves are 1 trillion trillion trillion times as weak as electromagnetic waves. Detecting and measuring gravitation waves is, therefore, a difficult problem for the modern physicist.

When electromagnetic waves hit matter, they shake just the charged particles (protons and electrons). When gravitational waves hit matter, all particles are shaken. The sizes of the vibrations are vastly different: electromagnetic waves shake particles a trillion trillion trillion times as vigorously as gravitational waves. In addition, the mode of vibration is quite different in the two cases. Consider a cloud of charged particles in space. As an electromagnetic wave passes through the cloud, all the charges oscillate in unison back and forth in the same direction. The direction of oscillation is perpendicular to the direction in which the electromagnetic wave is traveling. Now imagine a similar cloud of particles in space being traversed by a gravitational wave. As before, the direction of oscillation is perpendicular to the direction of propagation of the wave. But, unlike the way in which the particles in the electromagnetic case behaved, the individual particles are now set in motion relative to each other. If you were sitting on one of these particles, at one instant you would notice that the particles to your left and right were moving away from you, while the particles above and below were coming toward you. An instant later, this situation would be reversed: the particles on the left and right would be approaching you, while the particles above and below would be receding from you. These two contrasting cases are shown in Figure 10-1.

This mode of vibration clearly suggests a variety of means of detecting gravitational waves. Suppose instead of a cloud of particles you have a large object, such as a metal cylinder. Every time a gravitational wave passes through the cylinder, the cylinder will flex in response to the wave. The only remaining problem is inventing the technology that will be able to detect the minute changes in the shape of the cylinder.

Many of the monumental problems associated with building antennas for the detection of gravitational waves were solved during the 1960s by Dr. Joseph Weber of the University of Mary-

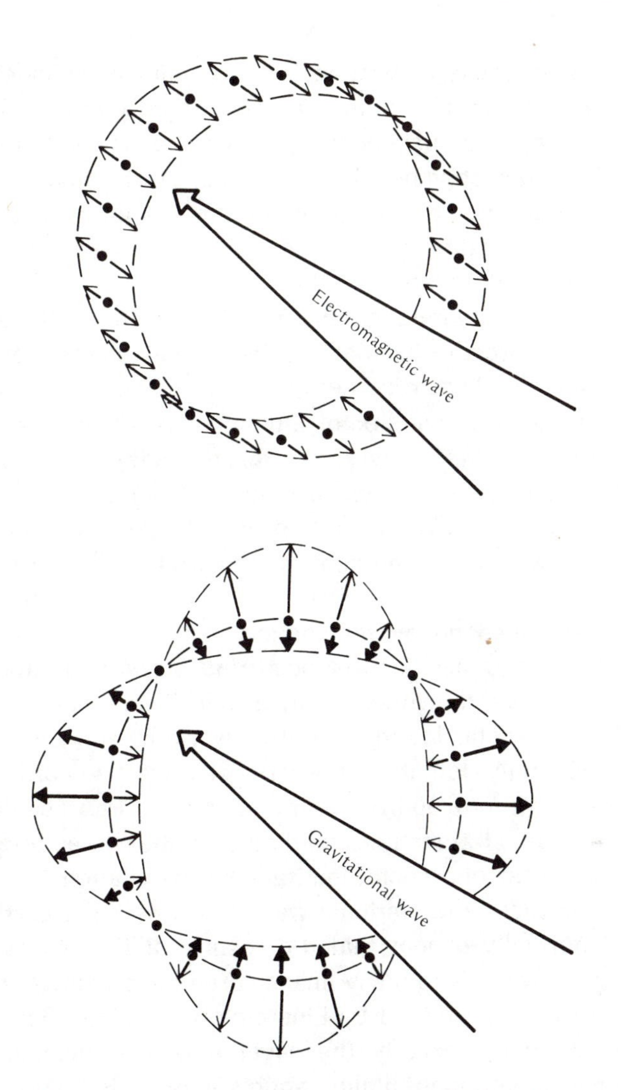

Figure 10-1. Electromagnetic and gravitational waves. Consider a ring of elctrons suspended in empty space. If an electromagnetic wave (light, radio wave, and so on) strikes this ring of electrons, all the electrons vibrate back and forth in unison. If a gravitational wave strikes this same ring of particles, the electrons are set in motion relative to each other.

land. Dr. Weber's antennas consist of large aluminium cylinders two feet in diameter, five feet long, weighing slightly more than one ton. Such a cylinder is suspended by a wire around its middle and hung in a vacuum chamber. This chamber, furthermore, is isolated from the outside world by an elaborate system of shock absorbers. When a gravitational wave passes through the cylinder, stresses are set up. To detect the resulting oscillations, Dr. Weber mounted piezoelectric crystals on the surface of the cylinder. These crystals translate the oscillations into small electric currents, which are amplified and recorded.

Unfortunately, such a cylinder of aluminum is always oscillating due to thermal effects. To overcome this difficulty, electronic filters were installed in the system to ignore all but the largest oscillations. In addition, Dr. Weber set up two of these antennas, one at the University of Maryland near Washington, D.C., and the other at the Argonne National Laboratory outside Chicago. These two antennas are linked up by telephone lines in such a way that simultaneous large oscillations occurring at both stations are readily recorded, as shown schematically in Figure 10-2.

In 1969, Dr. Weber startled the scientific world by announcing that he had successfully detected gravitational waves. According to his calculations, the probability of simultaneous, large oscillations occurring at both stations is exceedingly small. Yet, as often as once a day, at least one such simultaneous oscillation is recorded, suggesting that a gravitational wave has struck the earth.

One would normally expect that the results of Dr. Weber's experiment would have solved many unanswered questions. After all, gravitational waves predicted by Einstein's theory had finally been observed. Actually, exactly the reverse is the case. Dr. Weber's equipment, although highly sophisticated, is actually quite crude. Such equipment could not possibly detect gravitational waves due to any ordinary astronomical events. Even extraordinary violent events could not be expected to occur as often as once a day. If Dr. Weber's measurements are reliable—and some scientists question this—then what could he possibly be observing?

Weber's instruments suggest that the gravitational waves are coming from the center of our own galaxy, some 30,000 light years from the earth. If this is correct, then each burst observed by Weber must be the result of an extremely violent event. Calcu-

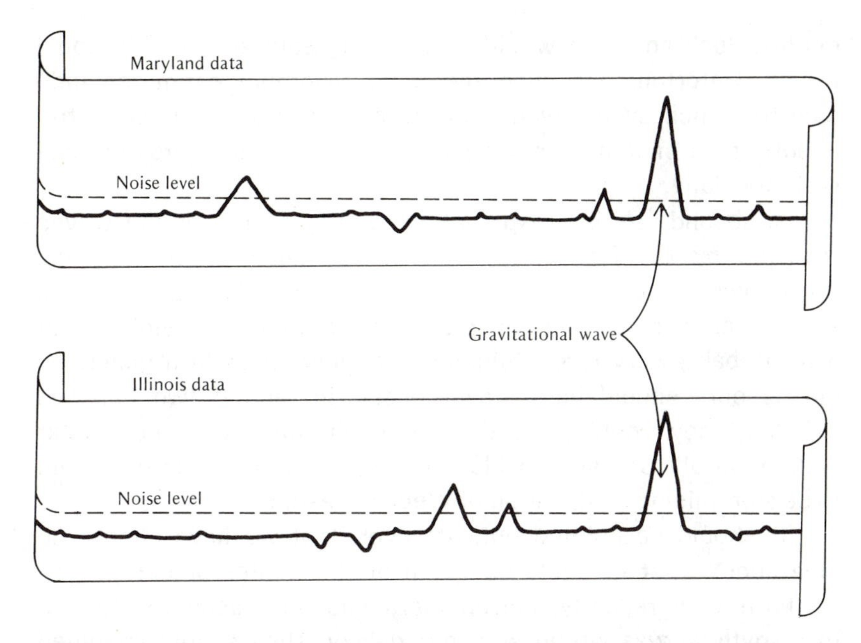

Figure 10-2. Data for the detection of gravitational waves. The aluminum cylinders used by Weber are always vibrating for one reason or another. All vibrations that are below the "noise level" are ignored. If two strong signals occur at the same time at two different receiving stations, then it is argued that we have observed a gravitational wave.

lations reveal that the energy in each of these events must be equivalent to one to ten solar masses. In other words, on the average, the entire mass of a star several times the size of the sun must be converted almost instantaneously into energy to give rise to each burst of gravitational radiation observed by Weber.

There are only two types of events that could produce this amount of energy. The first of these explanations deals with the collapse of a massive star to form a black hole. As described in earlier chapters, when a massive star has exhausted all its nuclear fuel, there are no longer any sources of energy to support the star. The star then begins to collapse rapidly under its own weight. Eventually, when the entire massive star has shrunk to a size only a few miles in diameter, the gravitational field is so great that space-time folds in over itself and the star disappears from the universe. What is left is called a black hole. The formation of a black hole is a violent event involving many solar masses of matter. Such an event would be a strong source of gravitational waves. Unfortunately, even the most favorable estimates indicate that such events would not be expected to occur more frequently than once a year. Weber observes bursts of gravitational radiation daily.

A second possible explanation for the source of the waves reported by Dr. Weber involves the collision of black holes. If two black holes collide, they swallow each other up to form a slightly larger black hole. Such an event is extremely violent, but also probably very rare. Collisions between stars in a galaxy as old as ours should hardly ever occur. In fact, if two galaxies collided, they could pass through each other with no stellar collison at all. Collision of black holes, therefore, does not seem to be a promising explanation of Weber's results.

It should be pointed out, however, that we have almost no idea about what is really going on at the center of our galaxy. Up until very recently, astronomers had no reason to suspect that anything was wrong with our galaxy. They simply imagined that the galactic nucleus consisted of lots of ordinary stars. Recent observations have dramatically altered these opinions. For example, Dr. Frank Low of the University of Arizona has discovered three exceedingly bright infrared objects in the constellation of Sagittarius in the direction of the center of our galaxy. These objects, which had been unknown before observations with infrared telescopes, are responsible for almost 10 percent of the total energy output of our entire galaxy! In addition, there is mounting evidence that our entire galaxy is in the process of exploding! This explosion is quite mild compared to M82 or NGC 1275, but nevertheless we are now beginning to appreciate how little we know and understand even about our own galaxy.

Based on current data and the best estimates, astrophysicists simply cannot account for Dr. Weber's daily observations of gravitational waves. This has led many scientists to suspect Weber's measurements, although no one has been able to demonstrate how Weber's results might simply be wrong. Perhaps we are just not aware of what is going on at the center of our galaxy, and we must wait for new and exciting discoveries about events and processes, which, at present, are even beyond our imagination.

An important advance in the field of gravitational radiation would be the confirmation of Weber's results by an independent team of scientists. They could repeat Weber's work with identical experiments or use totally different methods. For example, if a gravitational wave passes through our solar system, then the moon should oscillate slightly back and forth relative to the earth. By placing mirrors on the surface of the moon and using lasers to measure the earth-moon distance very accurately, it should be possible to detect gravitational waves. Also, accurate measurements of the distances to interplanetary spacecrafts should serve the same purpose. There are numerous experiments, including the use of the Mössbauer effect suggested by me, which could be used as alternative methods for detecting gravitational waves. The only real drawback seems to be that all such experiments require enormous accuracy and painstaking precision, while current calculations predict that we probably will not observe anything at all.

Nevertheless, the confirmed detection of gravitational radiation during the late 1970s would mark one of the most dramatic advances in astronomy of this century. Such a discovery would be comparable to the detection of electromagnetic radiation by Hertz in 1888 and would furnish astronomers with a new and important tool for probing the universe.

11

The Shape of the Universe

One of the most fascinating subjects in astronomy deals with the properties and evolution of the universe as a whole. As we saw in Chapter 5, an important tool in this regard is the redshifts observed in the spectra of distant galaxies. Nearby galaxies have their spectral lines shifted slightly toward the red end of the spectrum, while the spectral lines of more distant galaxies are shifted by substantial amounts toward the red. From these data astronomers conclude that nearby galaxies are moving away from us slowly, while the more distant galaxies are receding from us more rapidly. If the speed of recession of galaxies is plotted against their distance, we obtain a graph such as that shown in Figure 11-1.

The diagram, which visually displays the so-called Hubble law, is perhaps one of the most important graphs in astronomy. From the observation that the distance and speed of galaxies are related in this simple linear fashion, astronomers immediately concluded that the universe as a whole is expanding! The inclination or slope of the line in Figure 11-1 tells us the rate of expansion.

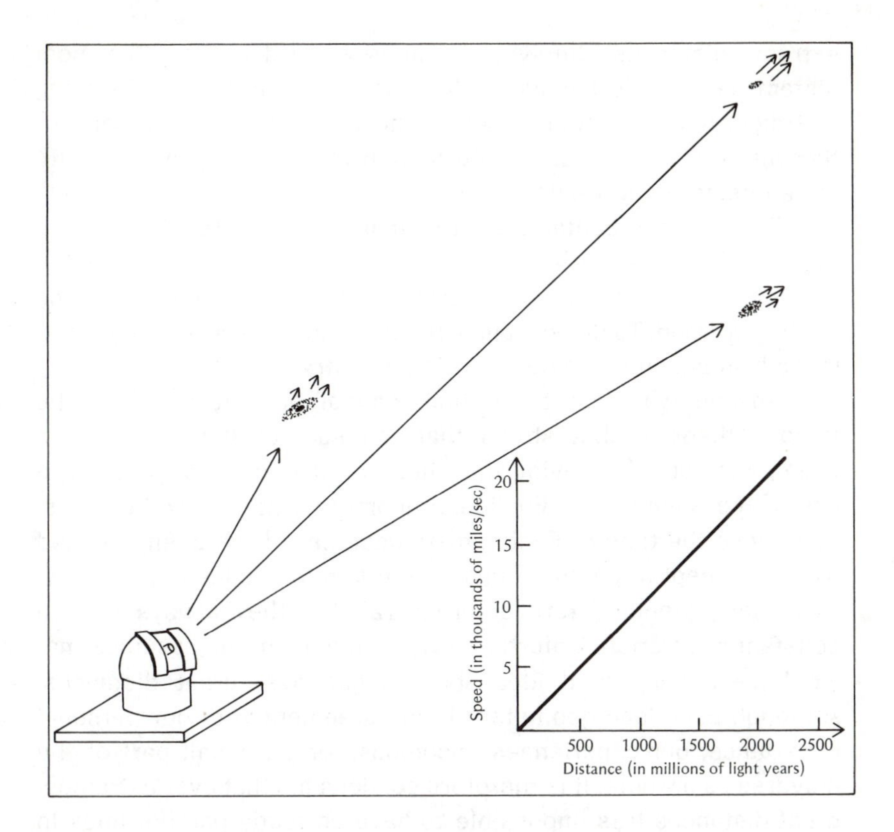

Figure 11-1. The Hubble law. The more distant a galaxy is, the more rapidly it is receding from us. The distances and recessional velocities of galaxies are related in a very simple linear fashion.

In this regard, it is interesting to note that when Dr. Einstein first tackled the problem of the structure of the universe from the viewpoint of general relativity, he discovered that the universe should indeed be expanding. Unfortunately, the Hubble law was unknown at that time, and Einstein doubted his calculations. In order to eliminate universal expansion, he added a "cosmological constant" to his field equations and thereby missed the chance to predict one of the most important astronomical discoveries of this century.

Once the expansion of the universe had been verified and accepted, astronomers found themselves faced with a variety of new and fascinating questions. The data from which Figure 11-1 was drawn are based on observations of galaxies that are not at extreme distances. But what happens when data from the most distant galaxies is included? Will the data continue to lie along a straight line? Questions such as these are extremely important; the answers give detailed information about the shape and ultimate fate of the entire universe.

The universe contains a lot of matter and energy in the form of galaxies, quasars, light, radio waves, and so on. Due to the presence of this matter, the space-time of the universe will be slightly warped. To understand what is meant by a curved universe, it is advantageous to review a little geometry.

Anyone who has taken high school geometry recalls the famous theorem that states that the sum of the angles of a triangle is 180°. In proving this theorem, it is necessary to make use of the postulate of Euclidean geometery about parallel lines. Ever since the time of Euclid, mathematicians have been troubled by the concept of parallel lines. According to Euclidean geometry, two lines extending forever are parallel if they always have a constant separation. Unfortunately, such a definition assumes that we have a very good idea about space over great distances. Although Euclidean geometery is in agreement with our personal experience, our experiences encompass only a small part of the universe as a whole. It is therefore conceivable that over extremely great distances it is impossible to have perfectly parallel lines in the strict Euclidean sense.

There are three possibilities. First of all, perhaps Euclidean geometery is correct. Perhaps, if we were to send two idealized beams of light out in the same direction separated by a certain distance, even billions of light years from our solar system, we would find these same two beams traveling alongside each other, just as when they started off. In such a case, we would say that space is *flat* and that the geometry of space is Euclidean. In such a case, if we were to draw a gigantic triangle with our solar system at one corner and two distant galaxies at the other corners and if we were to measure the angles at the corners of this huge triangle, we would find that the sum of these three angles is exactly 180°.

The second possible geometry of space is that all lines that start off being "parallel" must intersect somewhere. In other words, if we were to perform our experiment with two powerful searchlights pointed in the same direction, we would find that at some distance from our solar system these two beams crossed each other. This would happen even if we took great care in aligning the searchlights so that the beams at least started off being "parallel." In such a case, space is non-Euclidean, or curved. We would then say that space is *positively curved,* such as the surface of a sphere. In that case, if we were to draw a gigantic triangle with our solar system at one corner and two distant galaxies at the other corners and if we were to measure the angles at the corners of this huge triangle, we would find that the sum of these three angles is greater than 180°.

Finally, the third possibility is that two beams of light that start off being "parallel" must always diverge. In other words, far from our solar system the two beams of light from our searchlights are getting farther and farther apart, regardless of how carefully we set up our experiment. We would then say that space is *negatively curved,* such as on the surface of a saddle. With regard to our gigantic triangle, we would find that the sum of the angles of this or any other such triangle would be less than 180°. These three cases are shown schematically in Figure 11-2.

Clearly, one of the most significant questions we could ask is simply: Which of the three possible geometrics is correct for our universe? To tackle this problem, we must first make an important assumption. As we look far out into space, we find great numbers of galaxies. Most of these galaxies occur in clusters that are distributed more or less at random. For the purposes of determining the overall curvature of the universe, we shall, therefore, assume that *galaxies are distributed at random* throughout space. In other words, on the average, the number of galaxies in one part of the universe is the same as the number of galaxies in any other part of the universe. This is true, of course, only if we consider very large regions of space.

In order to derive a method for distinguishing the actual curvature of space, consider three universes corresponding to the three possible geometries in which galaxies are randomly distributed. To picture this, imagine taking three surfaces: one flat, one positively curved, and one negatively curved. Paint each of the surfaces with glue, and sprinkle salt over each. Let each grain of salt represent a galaxy or cluster of galaxies in each of the possible universes. These three cases are shown schematically at the top of Figure 11-3.

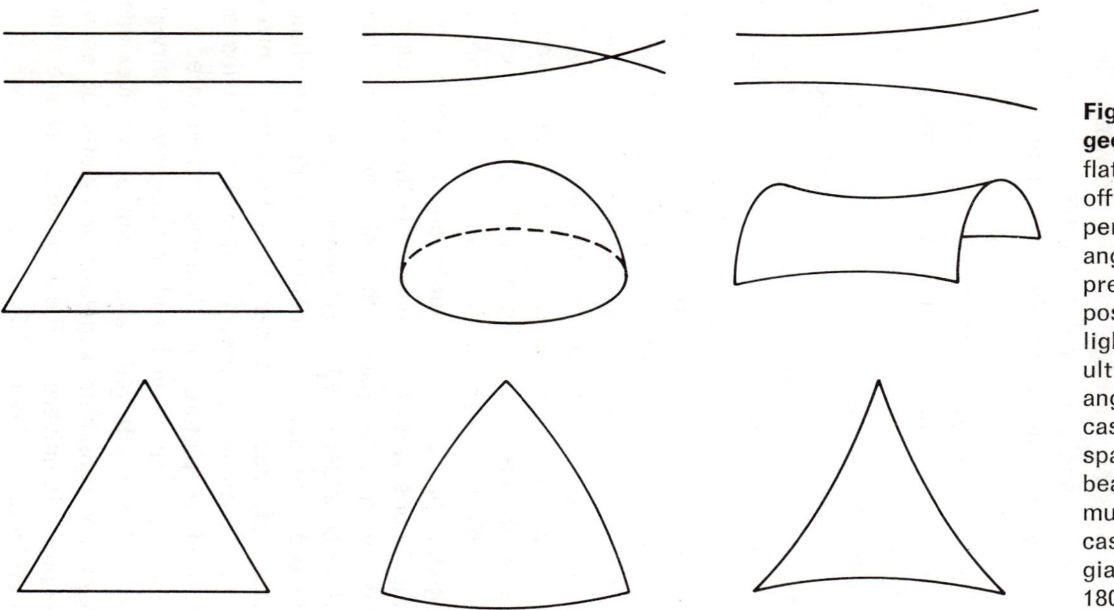

Figure 11-2. The three possible geometries of space. If space is flat, then beams of light that start off parallel will forever remain perfectly parallel. The sum of the angles of a giant triangle will be precisely 180°. If space is positively curved, then beams of light that start off parallel will ultimately cross. The sum of the angles of a giant triangle in this case will be greater than 180°. If space is negatively curved, then beams of light that start off parallel must ultimately diverge. In this case, the sum of the angles of a giant triangle will be less than 180°.

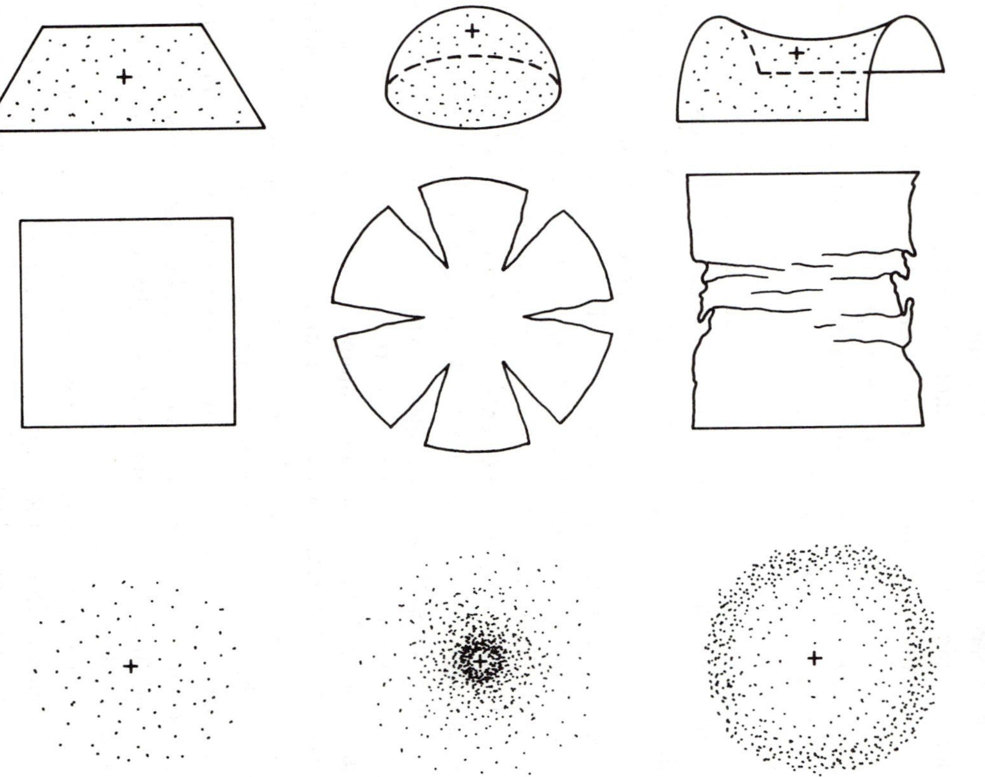

Figure 11-3. Galaxy counts and the curvature of space. Galaxies are randomly distributed in space. If space is flat, when we plot the locations of galaxies on a piece of paper we will find that the points are evenly distributed over the paper. If space is positively curved, then when we plot the locations of galaxies we will find that there seem to be more galaxies nearby than far away. Finally, if space is negatively curved, the plot would suggest that there are excess numbers of galaxies at great distances.

If astronomers are located on one of these three universes, they have, offhand, no way of knowing which way their universe is curved. At best, all they can do is look out into space and measure the distances to the various galaxies or clusters of galaxies. At this point, however, we notice something very unique. If their universe is flat, when they plot, on a flat piece of paper, the locations of the objects they observe, they will find that the objects are randomly located. If their universe is positively curved, after plotting the results of their observations on a flat piece of paper, they will find that there seem to be more galaxies nearby than far away. And, finally, if their universe is negatively curved, after plotting their results on a flat piece of paper, they will find that there are more galaxies at great distances compared to nearby. The reason for this is that the astronomers, by making a map on a flat piece of paper, are forcing their data onto a flat surface.

In other words, if space is flat, we should find that the number of galaxies per unit volume is essentially independent of distance. If space is positively curved, such as the surface of a sphere, then the number of galaxies per unit volume should decrease with increasing distance. And if space is negatively curved, such as the surface of a saddle, then we should find an increasing number of galaxies per unit volume with increasing distance.

Unfortunately, these effects are extremely small and become pronounced only at very great distances. If space is curved, the curvature must be so slight that, near our own galaxy, space looks flat. Consequently, astronomers are interested in the number of astronomical objects per unit volume very far from our solar system. But the farther away an object is, the dimmer it appears. Optical telescopes cannot "see" very far into space, and, therefore, observations even from the 200-inch telescope at Mt. Palomar are of little value in helping us determine the curvature of the universe. Only recently have radio telescopes been constructed that have the necessary sensitivity to detect extremely distant radio sources.

Looking out into space with sensitive radio telescopes, the astronomers detect thousands upon thousands of faint extragalactic sources. The 4C and 5C catalogs (successors to the famous *Third Cambridge Catalogue)*, as well as the monumental *Ohio Survey* recently completed at Ohio State University, list an incredible number of radio sources that might shed some light on the true geometry of the universe. Unfortunately, nature is more complicated than our simple example with grains of salt in Figure 11-3.

In order to determine the geometry of the universe, astronomers must focus their attention on the most distant objects. But the most distant objects are also the faintest. Most of the faint radio sources do not correspond to any visible object. Astronomers therefore do not have a direct method of determining the distances to individual faint radio sources. Consequently, they are not capable of drawing a map similar to those shown at the bottom of Figure 11-3. An alternative approach must be adopted.

Suppose we assume that *all* extragalactic radio sources are intrinsically alike. Ideally this is like saying "all automobile headlights are the same." Of course, there are some that are brighter than the average and some that are dimmer. But *if* all extragalactic radio sources have essentially the same intrinsic luminosity, then their observed brightness (i.e., the signal strength detected by the radio astronomer) is a direct measure of the distance of the sources. If our assumption is correct, then the faint sources are far away and the bright sources are nearby. The fainter a source is, the greater is its distance.

This assumption concerning the relationship between the distance and strength of radio sources allows astrophysicists to circumvent the task of drawing a detailed map to determine the geometry of the universe. Instead, they simply *count* the number (N) of sources in various parts of the sky that have a certain signal strength (S) or greater. They expect to observe relatively few bright sources because these sources are presumably contained in the small region of the universe near our galaxy. Moving to greater and greater distances, however, their observations encompass larger and larger volumes of the universe. Consequently, the number of observed sources should increase as the radio astronomers search for fainter and fainter objects. Assuming that the universe is flat, there is a simple relation between the source count number (N) and the signal strength (S), as shown in Figure 11-4. Ideally, deviations from the straight line in this graph should reveal deviations from flatness in the curvature of the universe. An excess number of bright (i.e., nearby) sources would suggest a positively curved universe while an excess

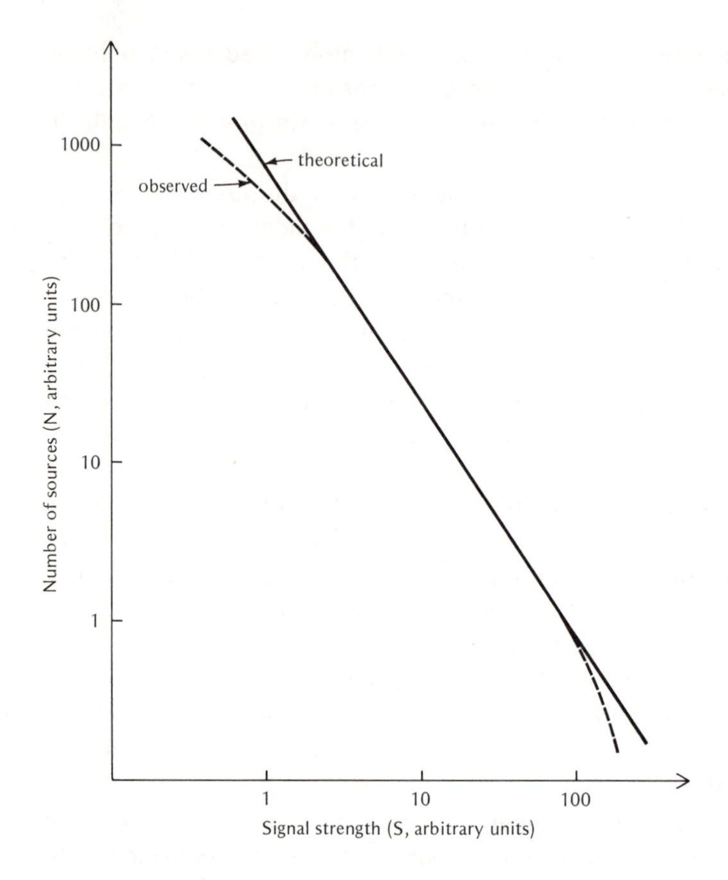

Figure 11-4. The log N–log S plot. In principle, it should be possible to deduce the geometry of the universe by plotting the number (N) of extragalactic radio sources against their signal strengths (S). The "theoretical" line is for flat space. The "observed" curve suggests that evolutionary effects cloud the picture.

number of dim (i.e., distant) sources would indicate a negatively curved universe.

Unfortunately, the data (dashed curve in Figure 11-4) do not conform with any of our expectations. The reason seems to be associated with the *evolution* of radio sources. Recall that as you look out into space, you are also looking backwards in time. When astronomers observe a radio galaxy one billion light years away, they are receiving data that is one billion years old. Observing still more distant sources, astronomers are seeing what radio galaxies looked like further and further into the past. It seems entirely reasonable to suppose that radio galaxies change with time, and, consequently, the same class of objects seen at various distances from earth could have very different stages of their life cycles. In other words, our assumption that "all extragalactic radio sources are alike" is faulty. Nevertheless, we notice that the data in Figure 11-4 do not differ very greatly from the Euclidean flat-space curve.

Perhaps the problem of the shape of the universe could be tackled from an entirely different approach. Recall that the distribution of matter in the universe is responsible for the geometry of space-time according to general relativity. The gravitational fields of the galaxies scattered throughout the universe directly influence the curvature of space. But obviously all these galaxies exert a gravitational force on each other. The mutual gravitational attraction of all the galaxies, therefore, tries to *slow down* the overall expansion of the universe. The expansion of the

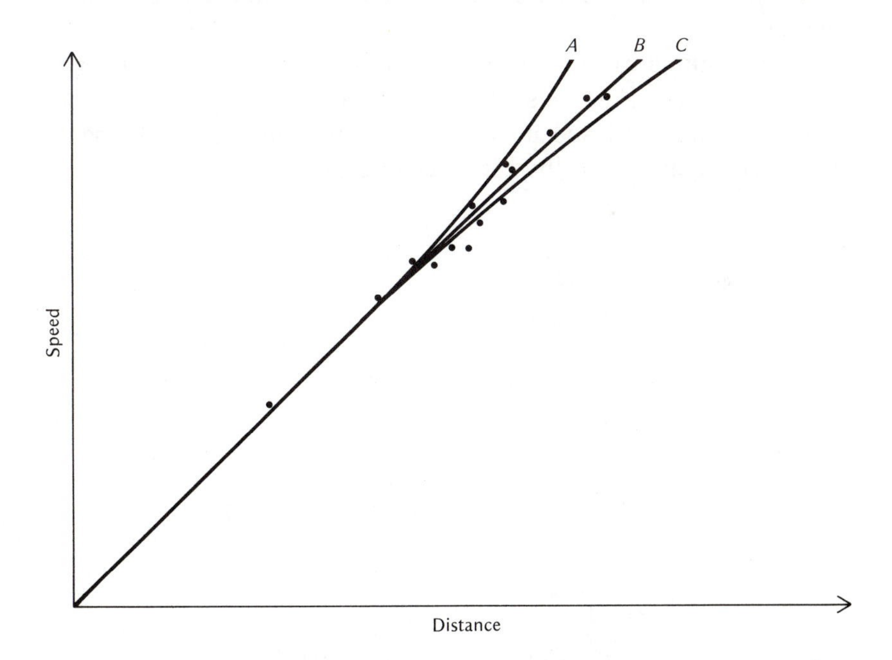

Figure 11-5. The Hubble law extended to great distances. The shape of the universe affects how the Hubble law looks when we include data for the most distant galaxies. Curve *A* corresponds to a positively curved universe, curve *B* to a flat universe, and curve *C* to a negatively curved universe. The data seem to favor curve *B*.

universe must be decelerating. The question is, How rapidly is the expansion of the universe slowing down? This is equivalent to asking, How does the Hubble law behave for the most distant galaxies? By examining the distances and speeds of the most remote galaxies, astronomers should be able to measure the deceleration of the expansion of the universe.

Figure 11-5 shows three possible extensions of the Hubble law to great distances. Curve *A* corresponds to the case of rapid deceleration. If there is enough matter in the universe to cause a substantial deceleration, then there is enough matter to cause the universe to be positively curved. Consequently, curve *A* corresponds to a positive curvature of space. Curve *B* denotes a more gradual deceleration and corresponds to a flat geometery of space. Curve *C* stands for a very slight deceleration and corresponds to a negative curvature of space. Therefore, from observing the degree of deceleration in the expansion of the universe, it should be possible to deduce the shape of the universe.

Observations during the mid-1970s by Drs. A. Sandage and G. A. Tammann at Hale Observatories indicate that the data favor curve *B*. As we shall see in the next chapter, this has important implications concerning the ultimate fate of the universe.

12

The Fate of the Universe

One of the most important discoveries in astronomy during the past century was the realization that the universe is expanding. As we have seen, this hypothesis came to light as a result of the fact that the distances to galaxies are correlated with the redshifts observed in their spectra. The more distant a galaxy is, the faster it is receding from us. The problem that then faces cosmologists is how this information can be used to formulate ideas about the nature and properties of our universe in the distant past and distant future. With regard to this matter, two prevalent theories have emerged that shall now be treated in a qualitative manner.

Because the universe is expanding, it is obvious that the clusters of galaxies astronomers observe in the sky are getting farther and farther apart. The simplest and most straightforward idea that comes to mind is that if we could see what the universe looked like many years ago, we would find the galaxies more closely crowded together. Indeed, at some time in the very distant past, all the matter in the universe must have been packed together, producing an extremely high density. Such a situation is reminiscent of the condition of matter inside a black hole. Presumably, therefore, at that time in the distant past, there must have been a *"big bang"* that caused the universe to begin expanding. The theory that speaks of a primordial explosion is therefore called the *big bang theory.*

Several decades ago, an alternative theory was proposed by H. Bondi, T. Gold, and F. Hoyle. No one argued with the idea that the universe is expanding. But they hypothesized that perhaps, as the distances between clusters of galaxies increase, new galaxies are created to fill the resulting void. In other words, if we were to move into the distant future, we would find that the universe looks roughly the same as it does now. Of course, many details would be quite different. There would be new people walking around, there would be new stars in the sky. But the number of galaxies in a cubic billion light years, for example, would be essentially the same as it is today.

Conversely, in the distant past we would also have had approximately the same density of galaxies and matter in the universe as we do today. Galaxies are not required to be crowded on top of each other as in the early stages on the big bang theory, simply because many of the galaxies we see today had not as yet been created. As a result, the universe is in a *steady state.* The theory that speaks of a "continuous creation" of matter (by some mysterious, unknown process) is therefore called the *steady state theory.* These two theories are contrasted schematically in Figure 12-1.

During the 1960s, a great deal of evidence came to light that severely discredited the steady state theory. First of all, from the Hubble law, astronomers find that the big bang should have occurred at least 15 billion years ago. This date is obtained simply by extrapolating backward to a time when all the galaxies should have been packed together. Second, the oldest star clusters seen in the visible universe also seem to have an age of about 15 billion years. And finally, from the study of isotopes (e.g., uranium-lead content), scientists again came up with an age of around 15 billion years. The approximate agreement of these ages is very striking.

A second major blow was dealt to the steady state theory with the discovery of the so-called microwave blackbody background radiation by A. A. Penzias and R. W. Wilson at Bell Labs in New Jersey in 1965. Penzias and Wilson were involved in the

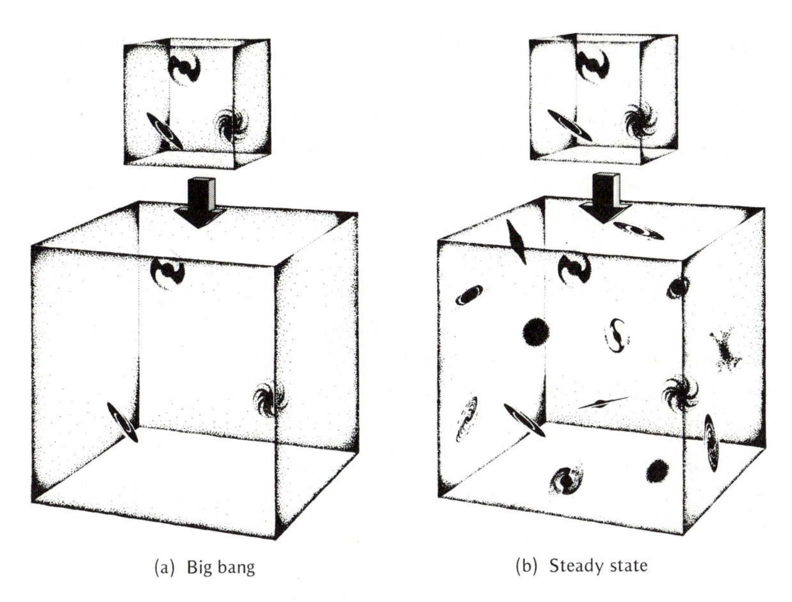

Figure 12-1. Big bang theory versus steady state theory. This schematic diagram shows two views of a section of our universe at different times. If the big bang theory is correct, then the galaxies simply get farther and farther apart. If the steady state theory is correct, then as older galaxies separate, new galaxies are created to fill the resulting void.

construction of a sensitive horn antenna to be used primarily in conjunction with communication satellites. They were initially very annoyed to find a constant, steady, background noise, or static, whenever they pointed their instruments to the sky. Meanwhile, at Princeton Unviersity, Dr. Robert H. Dicke theoretically predicted that if the big bang theory was correct, then it should be possible to detect some of the radiation left over from the initial explosion, or *primordial fireball.* From detailed relativistic calculations, astrophysicists believe that a reasonable temperature of the universe one second after creation would be about 15 billion degrees Kelvin. As the universe expands, it also cools off, so that today the average temperature should be about 3° above absolute zero. An object at 3°K should radiate a small amount of energy in microwaves, primarily at wavelengths slightly longer than 1 millimeter. This corresponds very neatly to the background microwave noise that plagues the sensitive antennas at Bell Labs. In other words, if the theory is right, scientists are actually "seeing" the remains of the primordial fireball that created the universe.

It should be pointed out that one single measurement of the microwave background radiation does not verify the theory. Theoretically, the brightness of the background radiation should vary in a very specific fashion with wavelength, as shown in Figure 12-2. It was, therefore, necessary to make a series of observations at wavelengths from 100 centimeters down to about ½ millimeter for confirmation. Reliable data are now available and are in good agreement with a background temperature of 2.7°K.

Another important set of observations that seems to support the big bang theory is the discovery of quasars. As we noted in Chapter 9, up until 1960, most of the objects observed by radio

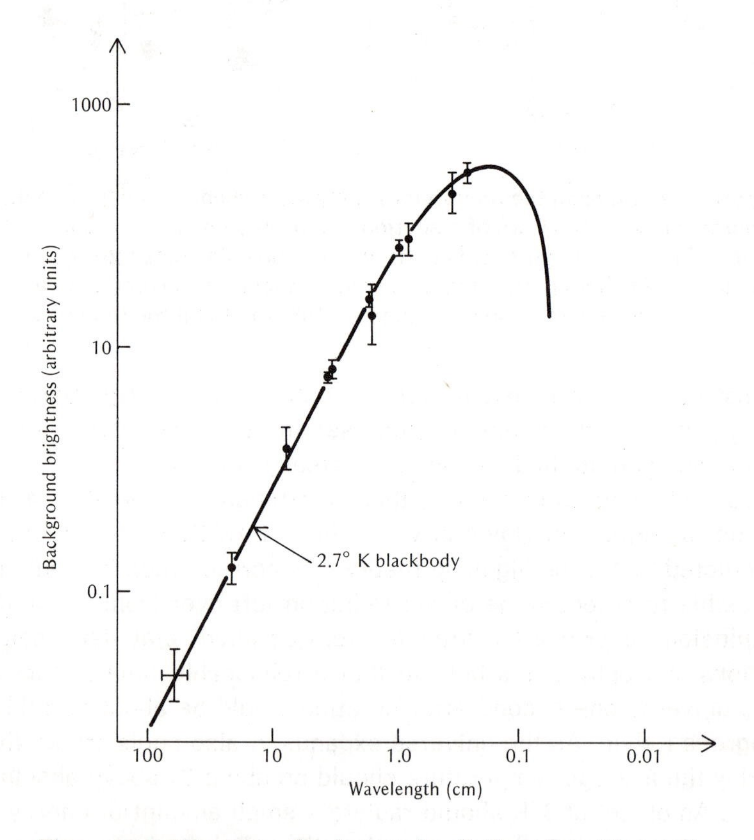

Figure 12-2. Microwave background radiation. Sensitive radio antennas detect microwave radiation from every part of the sky. This background radiation, which is uniformly distributed through space, corresponds to a temperature of 2.7° above absolute zero. Most scientists believe that this is the cooled-off "echo" of the big bang.

astronomers were large, diffuse sources. As the sensitivity and resolution of radio telescopes improved, however, astronomers focused their attention on certain sources that seemed to have unusually small angular diameters. In particular, the sources known as 3C 48, 3C 286, 3C 196, and 3C 147 were carefully observed. In October of 1960, Dr. Alan Sandage turned the 200-inch Palomar telescope toward the position for 3C 48 given by the radio astronomers. The only object found near this position looked like a bluish star. When he took a spectrum of this starlike object, to everyone's surprise he saw spectral lines that no one could identify. This mystery surrounding the spectrum of these so-called quasars was solved in 1963 by M. Schmidt, who found that quasars exhibited enormous redshifts. Indeed, some quasars have redshifts corresponding to speeds of up to 90 percent of the speed of light.

It is no exaggeration to state that quasars are among the most enigmatic objects ever discovered by humans. Accounting for their huge redshifts has plagued astronomers and astrophysicists for over a decade, and they seem no closer to a solution than they were fifteen years ago. As we have seen, there nevertheless appear to be essentially four possible approaches to explaining the redshifts:

1. gravitational redshift
2. Doppler shift
3. cosmological redshift
4. unknown physical laws

The first possibility was quite popular in the early 1960s. In fact, it was precisely this possibility that stimulated a revival of interest in relativity. The second possible explanation forms the basis of a school of thought initially proposed by Dr. Arp at the California Institute of Technology. Perhaps quasars are ejected with great violence from the centers of galaxies, and their redshifts reflect the enormous speeds with which they are traveling.

The third hypothesis, namely a cosmological interpretation, has been by far the most popular. In other words, astronomers turn distances and their redshifts result from the expansion of the universe. If this is so, then quasars should have profound cos-
to the Hubble law and conclude that quasars are at very great

mological implications. For example, quasars have large redshifts; there are no quasars with small redshifts. As we look out into space, we are actually looking backward into time, which, therefore, tells us that many quasars existed long ago, but that none are around today. If the cosmological interpretation is correct, these data would be a severe blow to the steady state theory. The universe must have looked very different several billion years ago with quasars all around. This would mean that the universe is evolving and not in a steady state.

In the early 1970s, a number of observations were made that seem to be in conflict with the cosmological interpretation. Although it may be too early to tell, some astrophysicists take the position that quasars are such an enigma we must consider the possiblity of new physical laws and effects as yet undiscovered. Among the possible candidates for new physical principles is a theory by F. Hoyle and J. V. Narlikar at Cambridge. They propose that certain properties of an object depend on the distribution of matter throughout the rest of the universe. As a result, the masses of atomic particles would have had different values in the distant past from those they have today. Objects would, therefore, have looked very different long ago from the way they look today.

In spite of some possible difficulties with the precise interpretation of the redshifts in the spectra of distant objects, the Hubble law can be used to tell a great deal about the future of the universe. We have seen that for most galaxies there is a simple linear relationship between distance and recessional velocity. Double the distance to a galaxy, and the speed with which it is moving away from you also doubles. But this is not necessarily so for the most distant galaxies. In observing the most remote objects, it should be possible to notice the effects of the mutual gravitational attraction between galaxies, which tends to slow down the expansion of the universe. As far as the familiar graph of the Hubble law is concerned, this means that the straight line bends either up or down near the end (see Figure 12-3).

If you stand on the earth and throw a rock up into the air, three things can happen. The rock could simply go up and come back down. If you throw the rock with great vigor (such as with the aid of a rocket), then the rock could escape from the earth's gravitational pull and fly off into space; at some extremely great distance from the earth, the rock would finally come to rest. The

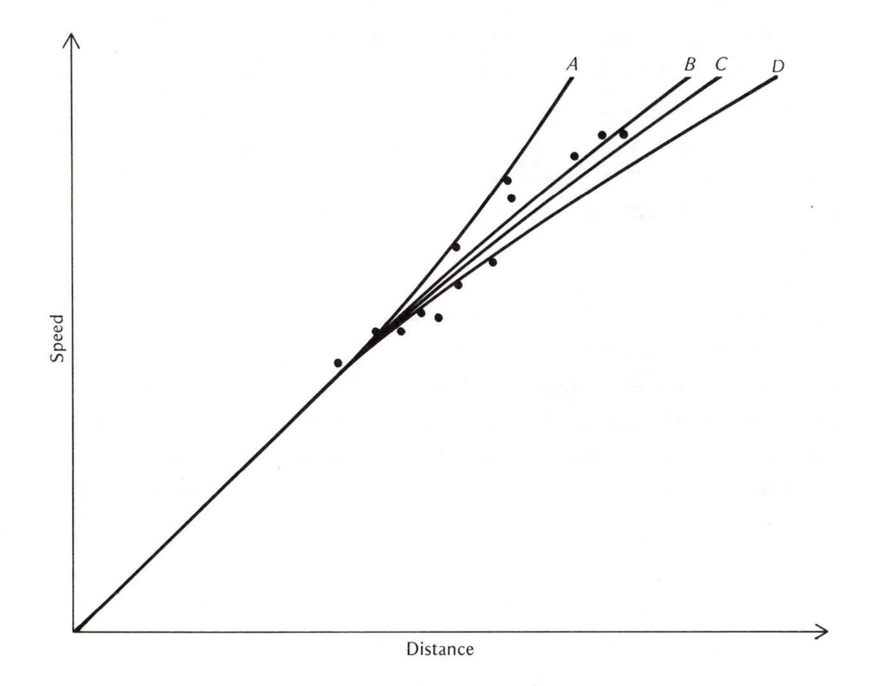

Figure 12-3. Behavior of the Hubble law at great distances. The theoretical graph of the Hubble relation for distant objects depends on the model of the universe used. Curves *A*, *B*, and *C* are for the big bang theory. Curve *D* is for the steady state theory.

third possibility is that with so much power expended in throwing this rock even infinitely far from the earth, the rock is still receding with a sizable velocity.

If the big bang theory is correct, a similar set of possibilities exists with the universe. If a comparatively small amount of energy was expended at the time of the big bang, then the galaxies will simply go up and then come back down. Although current observations indicate that all the galaxies are receding from us now, at some time in the distant future this recession will stop, and the galaxies will begin coming toward us as the universe begins to collapse in upon itself. The second possibility is that there was just enough energy in the primordial fireball so that when the galaxies are infinitely far apart, they will come to rest. In the case of the third possibility, there was so much energy in the big bang that the galaxies would continue to fly apart at high speeds forever.

These three possibilities are reflected in the behavior of the Hubble relation for distant objects. As shown in Figure 12-3, curves *A*, *B*, and *C* are the lines along which the data should lie if the big bang theory is correct. Curve *D* is for the steady state theory. In the case of curve *A*, the primordial fireball did not contain enough energy to separate the galaxies infinitely. At some time in the future, the universe will collapse in upon itself. In the case of curve *B*, when the galaxies are infinitely far part, they will finally come to rest. Curve *C* depicts the case wherein the primordial fireball was so violent that the galaxies will continue to fly apart forever. Figure 12-4 shows how the behavior of the Hubble law is related to the size of the universe as a function of time. As shown in Figure 12-3, the data seem to favor curve *B*, which corresponds to a universe that just barely manages to

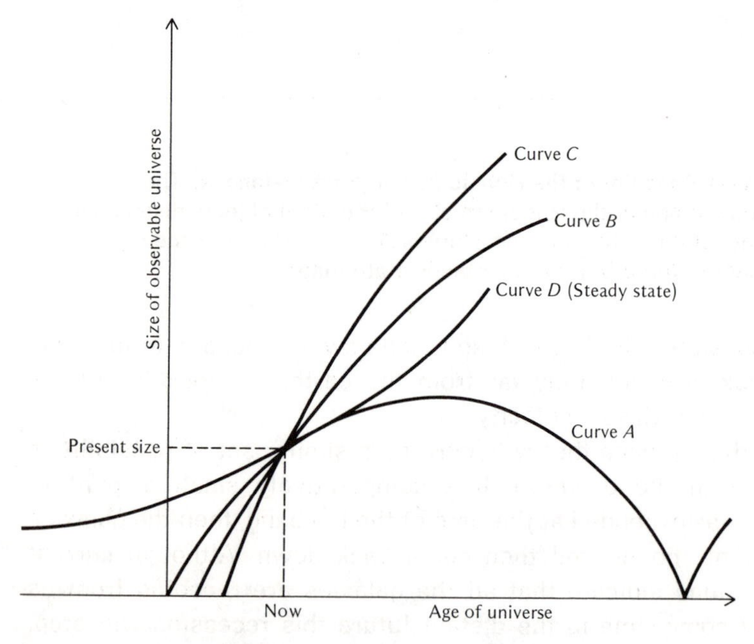

Figure 12-4. The extent of the observable universe. Specific theoretical models of the universe predict very different behaviors of the universe. If a relatively small amount of energy was expended in the big bang, then we live in an oscillating universe (curve *A*). Curve *B* corresponds to a big bang with just enough energy to ensure that the universe does not recollapse. If the big bang was extremely violent, then the universe will expand forever (curve *C*). Curve *D* is for the steady state cosmology, which has no "creation event."

expand forever. It should be clearly pointed out that these data contain large possible errors due to the extreme difficulty of the observations. Improving these data is among the most important tasks for astronomers in the 1980s.

Because the distribution of matter in space determines the curvature of space-time, we would expect that the expansion of the universe and the shape of the universe should be intimately related. This is indeed the case. As we saw in the preceding chapter, there are three possible geometries for space: hyperbolic, flat, and spherical. Curve *C*, depicting the most energetic expansion of the universe, corresponds to a negatively curved hyperbolic space. Curve *B* corresponds to a flat geometry. The oscillating universe of curve *A* has a positively curved spherical space.

Although all three curves (*A*, *B*, and *C*) are for a big bang cosmology, the three cases have quite different properties. If there is enough matter in the universe to stop the expansion, then there is enough matter in the universe to give space an overall positive curvature. The shape of the universe would then be spherical. Just as a basketball has a certain finite size, the universe would have a finite size. If the data favored curve *A*, we would be living in a *finite, closed* universe. Although the universe would have a specific size, this does not mean that it has an "edge" or a "center." If you put an intelligent bug on a basketball, that bug could walk around the basketball forever and never come to an edge or the center. The surface of a basketball has no edge and no center. It would therefore be totally meaningless to ask questions like, Where is the center of the universe? or What is beyond the edge of the universe or What is the universe expanding into?

In sharp contrast to a universe described by curve *A*, curves *B* and *C* correspond to universes that are *infinite* and *open.* If the universe is flat (curve *B*), or hyperbolic (curve *C*), then it extends forever in all directions. Such universes are infinite for all time, even at the instant of the big bang. Again it is meaningless to speak of an edge or a center; such terms cannot be applied to something that has an infinite extent. Although open universes extend forever, we can see only a finite amount of space. If the big bang occurred 15 billion years ago, then it is impossible to observe objects farther away than 15 billion light years. Even in open cosmologies, the "observable universe" has a finite size.

In open, infinite universes there is *one* big bang followed by endless expansion. But in a finite, closed universe the expansion ultimately ceases and contraction begins. Presumably this collapse terminates in yet another big bang, which creates a new universe from the ashes of the old. One of the exciting problems facing astrophysicists deals with questions relating to the nature of things before the last big bang and after the next big bang in an oscillating universe. In order to deal with these questions, we must return briefly to the subject of black holes.

When matter falls down a black hole left by a dead star, the matter loses all identity. As depicted schematically in Figure 12-5, if you drop a bathtub or a loaf of bread down a black hole, the end result is essentially the same. To be precise, the only quantities left after material falls down a black hole are the mass, charge, and angular momentum the object carried with it. Such things as the chemical composition of the material, the color of the object, and its size are all lost. Dr. John Wheeler of Princeton University has expressed this figuratively by saying, "Black holes have no hair." Nevertheless, the fundamental quantities of mass, charge, and angular momentum of an object falling down a black hole are retained because the black hole itself is attached to the rest of the universe. As shown in Figure 12-5, this is due to the fact that space is asymptotically flat far from the singularity. But if the entire universe itself falls down its own black hole, then even the fundamental quantities (mass, charge, angular momentum) are lost. This is because there is no flat space onto which we could connect our solutions to the field equations.

In examining the conditions of a universe when it collapses into its own black hole, Dr. Wheeler has found that a great number of important physical properties of matter become uncertain. For example, in our universe we find that matter has certain characteristics. Matter is made up of atoms, which consist of protons, electrons, and neutrons. Protons and electrons have equal, but opposite, electric charges. The mass of the proton, however, is 1836 times the mass of the electron. Any complete theory of the nature of physical reality must be able to explain this curious situation, as well as the reasons behind the peculiar number 1836. In addition, there are a great number of similar so-called dimensionless constants through physics that must also be accounted for. It now appears that when an entire universe col-

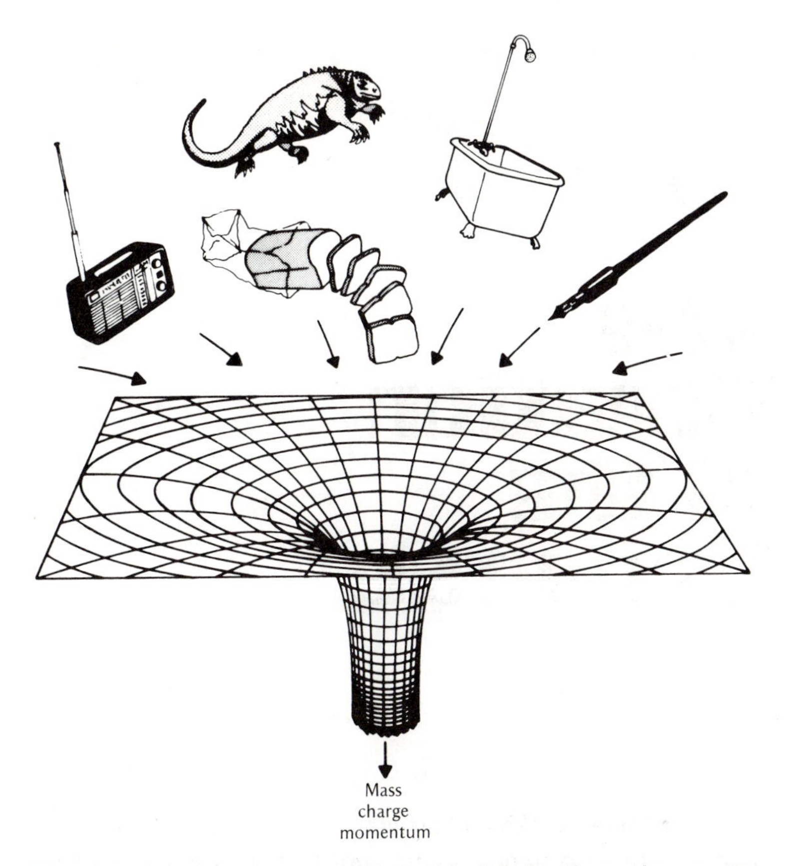

Figure 12-5. Material encountering a black hole. If you drop matter down an ordinary black hole, such as that left by a dead massive star, the final result is always essentially the same. Only the mass, electric charge, and angular momentum carried with the object are retained. If the entire universe goes down its own black hole, then even these fundamental physical quantities are lost.

lapses into its own singularity, the values of all these important quantities become uncertain. In other words, the universe that existed prior to the most recent big bang could have been fundamentally different from our universe today. Similarly, in the universe that will be created after the next big bang, the relative masses of nuclear particles, the electric charges they carry, and so on, could be quite different from those in our own universe. In other words, there seems to be a global uncertainty principle that applies to the basic, dimensionless constants at the time of universal collapse.

13

The Creation of the Universe

In many popular books and articles on relativity it was common to illustrate the meaning of the word *relativity* by appealing to familiar experiences in a train station. Imagine sitting on a train awaiting departure from a station. Looking out of the window, you see another train standing on the neighboring tracks. At the scheduled departure time, you again glance out of the window and notice that the neighboring train is moving past your line of sight. You settle back into your seat for an unexpectedly smooth ride. A moment later, however, you are shocked to realize that your train is still standing at the station. The motion of the neighboring train, which departed a few minutes earlier, gave you the illusion that you had already started on your journey. Only after the neighboring train ceased to fill your field of view could you appreciate what actually happened. The exposed railroad ties and stationary ground reveal that the other train was going in the opposite direction.

The usual purpose of this little story is to illustrate the fact that all motion is relative. It is meaningful to speak of motion only in relation to the locations of other objects in the universe. If the universe were totally empty, it would be impossible to talk about motion. One says that one is at rest or moving with a certain speed only because there are objects (the earth, the sun, our galaxy) in the universe against which a state of motion can be measured.

While the train station story has largely disappeared from the literature (perhaps because of the popularity of airplanes), a similar analogy can be used to illustrate some recent and revolutionary ideas in physics. Imagine sitting on an old-fashioned rotatable piano stool, as shown in Figure 13-1. By pushing against some nearby object, you can start yourself spinning. You know that you are rotating because you see familiar objects in the room moving past your line of sight. Your state of rotation is further verified when you move your arms in and out. Your hands experience centrifugal and Coriolis forces. Indeed, if you continued this bizarre experiment for any considerable time, these same forces acting on your inner ear and alimentary canal might cause you to become nauseous and dizzy.

Prior to the termination of the experiment, suppose you shut your eyes, and suppose that while your eyes were shut, the *entire* universe disappeared. Upon opening your eyes, you discover that you are alone in totally empty space. Even though you never stepped off the piano stool, *are you still rotating?* When you move your arms in and out, will your hands still experience centrifugal

Figure 13-1. A man on a piano stool. By considering the implications of rotation in an empty universe, it seems reasonable to suppose that the masses of atoms depend on the distribution of matter in space.

and Coriolis forces? Since there is nothing else in the universe, is it even meaningful to speak of "rotation" at all?

In classical physics, the sizes of centripetal and Coriolis forces are directly related to the inertial mass of the objects on which such forces act. Inertial mass is that property of matter that, according to Newton, resists changes in a state of motion. Yet, a state of motion can be intelligently defined only in relation to other objects in the universe. It, therefore, seems reasonable to suppose that mass is somehow related to the distribution of matter elsewhere in the universe. This conjecture, that the mass of an object is determined by the distribution of matter in the distant universe, is called *Mach's principle.* From this viewpoint, all of the above paradoxical questions has a very straightforward answer. After shutting your eyes, your hands in an empty universe would *not* experience the forces normally associated with rotation because the masses of the atoms in your body would go to zero as the outside universe vanished.

One of the most important advances in astronomy during this century occurred in 1929 as a result of the work of Edwin Hubble at Mount Wilson Observatory. In that year, Hubble announced his now-famous discovery that the redshifts of galaxies are directly proportional to their distances. Nearby galaxies have low redshifts, while more distant galaxies have much higher redshifts. It is customary to attribute these cosmological redshifts to the recessional motion of the galaxies. From the Doppler effect it therefore follows that nearby galaxies are moving away from us slowly while more distant galaxies are rushing away from us at much higher speeds. As we saw in Chapter 5, the final conclusion drawn from this straightforward line of reasoning is that *the universe is expanding.*

Nothing could seem more obvious to the modern astronomer than the idea that the universe is expanding. Very few astronomers contest this conclusion, and it is generally agreed that a primordial explosion (i.e., the "big bang") occurred some 15 billion years ago, which started the universal expansion.

It is instructive to contemplate the appearance of an expanding universe on a space-time diagram, as shown in Figure 13-2. As usual, time is measured vertically, while distance is measured horizontally. Furthermore, the space and time axes are scaled so that light rays travel along 45° lines. The bottom line in Figure

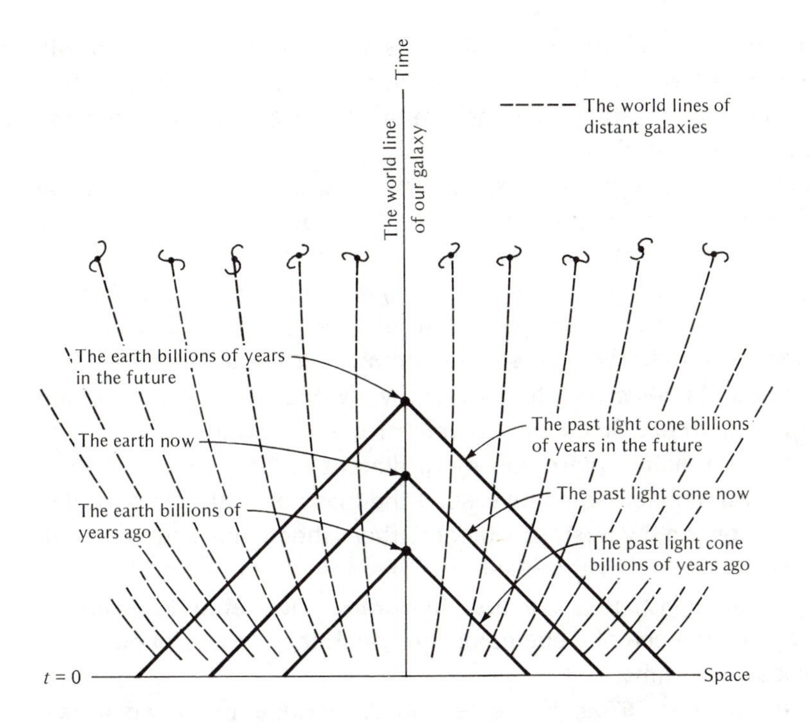

Figure 13-2. The expanding universe in space-time. At any instant, the extent of the observable universe is confined to the past light cone. With increasing time, the volume of the past light cone also increases.

13-2 (called "$t = 0$") denotes the time of the big bang, and the world lines of the galaxies gradually get farther and farther apart with increasing time, thereby reflecting the overall expansion of the universe.

In order to relate Figure 13-2 to the visible sky, remember that light rays travel along 45° lines in space-time diagrams. This means that galaxies observed by astronomers must lie along the *past light cone* extending downward from the location of our earth in Figure 13-2. Objects that are located outside of our past light cone lie in our "elsewhere" or "future" (refer to Chapter 2 and Figure 2-3) and, therefore, cannot as yet be observed.

It is important to realize that as time moves on, our past light cone encompasses a larger and larger volume of space-time. As shown schematically in Figure 13-2, billions of years ago—near the time of the big bang—the volume of our past light cone was comparatively small. It is now much bigger and will grow even larger billions of years in the future. This is simply a statement of the fact that the limits of the observable universe are increasing with time.

This picture of the expanding universe is entirely logical and self-consistent. But inherent in this logical self-consistency is the assumption that the basic properties of matter and the laws of physics are the same everywhere in space and for all time. Specifically, the expansion of the universe entails the assumption that the masses of atomic particles never change. Recalling Mach's principle, however, we wonder if this assumption is truly justified. Since the volume of our past light cone is constantly increasing, more and more distant objects fall within the limits of the observable universe. If the masses of atomic particles are influenced by distant objects, then these masses must *increase* with time as the number of distant objects along the past light cone increases. We are, therefore, motivated to examine the consequences of a cosmology in which the masses of particles increase with time.

In the mid-1970s, F. Hoyle and J. Narlikar proposed a revolutionary approach to cosmology. Arguing that the masses of particles (e.g., electrons, protons, and so on) are caused by interactions with distant matter in the universe, they succeeded in reformulating general relativity. Their equations look very much like Einstein's equations, except that the masses of particles are increasing with time.

To appreciate how the mass of an individual particle changes with time, consider the past light cone of that particle. As the particle gets older, the size of its past light cone gets bigger and bigger. As the volume of the past light cone increases, the particle can "see" more and more objects in the remote regions of the universe, just as in the case of the earth in Figure 13-2. Consequently, as the particle ages, it has an increasing number of *mass interactions* with these distant objects. As the number of interactions increases, the mass of the particle grows.

But notice that in the distant past, the light cone of a particle is very small and, therefore, its mass is low. In fact, at $t = 0$ (i.e., at the time of the big bang in traditional cosmology) the volume of a particle's past light cone is *zero.* Therefore, the mass of a particle at $t = 0$ is *zero.* In other words, in this new Hoyle-Narlikar cosmology, all objects (e.g., atoms, people, electrons, galaxies) start off with no mass at $t = 0$. As objects age, their masses increase. Today ($t =$ *15 billion years*) the masses of objects have the values observed in the world around us. Presumably, masses will continue to grow in the future.

This variation of mass permits a dramatically new interpretation of the Hubble law. Recall that the Hubble law says that there is a direct proportionality between the redshifts and distances of galaxies. But astronomers detect redshifts by measuring the displacement (toward longer wavelengths) of the spectral lines observed in the light from remote galaxies. Spectral lines, in turn, are caused by eletrons jumping between allowed orbits in atoms. In calculating the structure of atoms, physicists find that the wavelength of a particular spectral line is inversely proportional to the mass of the electrons orbiting atoms. This dependence of wavelength on the mass of electrons has been largely ignored by physicists. After all, the traditional viewpoint is that the masses of particles *never* change, and thus this relationship would have no bearing on any observations. But if, as suggested by Hoyle and Narlikar, the masses of electrons are increasing with time, then the wavelengths of spectral lines should be decreasing with time. Light emitted by ancient atoms (which have low-mass electrons) should have a longer wavelength than the same spectral lines emitted from modern atoms (which have higher-mass electrons). In observing distant galaxies, astronomers are seeing light emitted from atoms billions of years in the past. In other words, the redshifts of galaxies is *not* due to their recession but rather to the low mass of the electrons in the atoms of the galaxies!

Following this line of reasoning a little further, we see that the Hoyle-Narlikar approach permits a remarkable reinterpretation of the structure and behavior of the entire universe. Recall that the wavelength of a spectral line depends on the mass of the electron that jumps from one orbit to another. The reason for this is that the sizes of the allowed orbits around the nucleus depend on the mass of atomic particles. If the masses are low, then the orbits are spread out, resulting in long-wavelength (i.e., redshifted) spectral lines. As the masses of atomic particles increase, the sizes of the orbits shrink, and the amount of redshift becomes less and less. In other words, as shown schematically in

Figure 13-3, according to Hoyle and Narlikar the sizes of atoms are decreasing with time. Atoms shrink as they get older and older! But the sizes of atoms determine the sizes of everything. Atoms, people, rulers, galaxies, all must have been bigger in the distant past than they are today.

If you say that you are six feet tall, you mean that six 1-foot rulers could be placed end to end between your feet and your head. Similarly, when astronomers say that a galaxy is 1 billion light years away, they mean that exactly 1 billion rulers, each one light year in length, could fit between us and that galaxy. But if the lengths of rulers are shrinking with time, then rulers long ago must have been bigger than they are today. Consequently, fewer rulers would fit end-to-end between galaxies in the distant past. In other words, galaxies are *not* getting farther and farther apart. Instead, the units of length with which astronomers measure the distances to galaxies are getting shorter and shorter! The universe is *not* expanding! Instead, our rulers are shrinking!

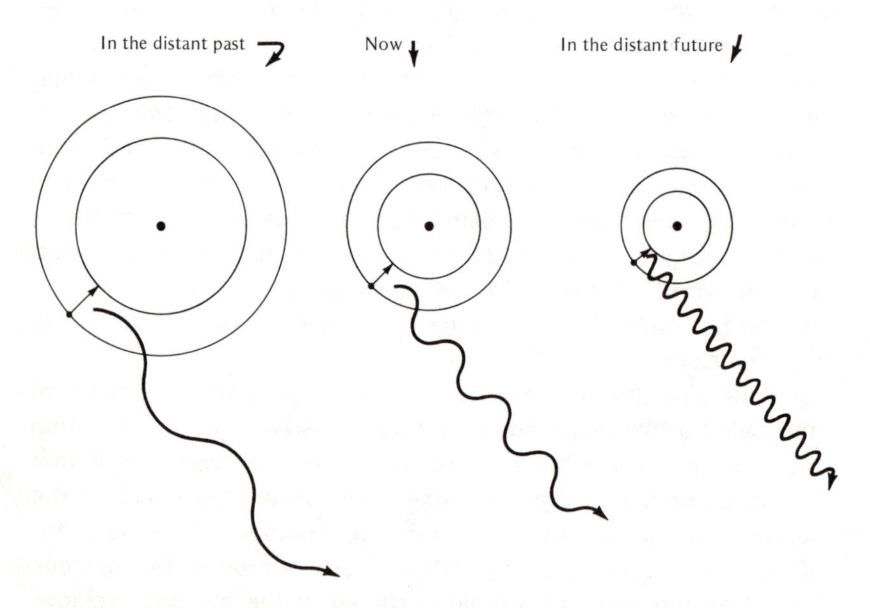

Figure 13-3. The sizes of atoms at different epochs. If the masses of electrons increase with time, then the sizes of orbits around the nuclei of atoms decrease. Large (low-mass) atoms emit spectral lines at long (redshifted) wavelengths. As the masses of particles increase, the sizes of atoms and the amount of redshift decrease.

The entire business amounts to a gigantic trade-off. The traditional view is that the masses of particles are the same for all time. Then, the redshifts of galaxies are due to their recessional speeds. The universe is expanding. There must have been a big bang. At the time of the big bang, all the matter in the universe was crowded together in a state of infinite density. *The big bang was a naked singularity* since no event horizons separate us from what happened 15 billion years ago.

In sharp contrast to this traditional view, Hoyle and Narlikar believe that the masses of particles increase with time. The redshifts of galaxies are due to the fact that atoms were bigger in the past than they are today. The universe is not expanding; instead rulers are shrinking. The expansion of the universe is an illusion. The big bang is an illusion. It never happened! Galaxies appear to have been crowded together 15 billion years ago only because they were very big in the distant past. Today they are much smaller and appear to be farther apart. In the spirit of T. S. Eliot, these revolutionary ideas of Hoyle and Narlikar could be called the *whimper cosmology.* Basic features of the big bang and the whimper are contrasted in Figure 13-4.

If these two cosmological views are indeed equivalent, then why bother with these strange ideas of Hoyle and Narlikar? There are perhaps a couple of reasons. First of all, over the course of the history of science, whenever a physicist's equations predicted "infinity" as an answer, the physicist simply had the *wrong* theory. For example, around the turn of the century, classical electromagnetic theory gave "infinity" as the answer in calculations involving radiation. The resolution of the difficulty was achieved with the invention of quantum mechanics, which gave finite results rather than infinite ones. Classical theory was simply the wrong approach, and our ideas had to be modified. Today, general relativity gives infinite answers. The big bang was a singularity wth infinite pressure, infinite density, and infinite curvature of space-time. Recent work by the brilliant British physicist Stephen W. Hawking strongly suggests that scientists should be very worried by a cosmology that has a naked singularity. We may just have the wrong theory. The Hoyle-Narlikar approach does not have a space-time singularity 15 billion years ago. The world lines of galaxies simply rise vertically in a space-time diagram of a whimper universe. The space-time geometry is very simple. Of

Figure 13-4. Big bang theory versus whimper theory. The traditional view is that the masses of particles never change; then atoms and rulers always had constant sizes and the universe is expanding. Conversely, if the masses of particles increase with time, then the sizes of atoms and rulers are shrinking and the universe is not expanding.

course, the price we pay for this geometric simplicity is that the masses of particles vary with time.

A second motivation that prompts us to consider the whimper cosmology seriously is related to the 3° blackbody background. In the ten years that followed the discovery of the microwave background by Wilson and Penzias (see Figure 12-2), a remarkable property of this radiation field was revealed. Regardless of the location toward which radio astronomers point their telescopes, they always observe *exactly* the same intensity of microwaves at a given wavelength. In other words, the 3° blackbody background is incredibly smooth, or *isotropic*. Assuming that this background is the cooled-off remains of the primordial fireball, astrophysicists find it extremely difficult to explain its high degree of isotropy. Indeed, all attempts to account for the smoothness of the radiation field from the framework of a big bang cosmology have met with only marginal success. The isotropy of the 3° background remains one of the great mysteries in modern astrophysics.

Consider the blue sky. The sky is blue because sunlight falling on the earth is bounced around or *scattered* by particles in the earth's atmosphere. This scattering is so effective that every part of the daytime sky (away from the sun and in the absence of clouds, of course) appears a uniform blue. The process of scattering of radiation has been very well understood for many years.

Now consider the whimper cosmology. As shown in Figure 13-5, the world lines of galaxies are vertical and parallel in a space-time diagram because the whimper universe is not expanding. But these world lines must have extended from times *earlier* than $t = 0$. Galaxies, planets, and stars could easily have existed at earlier times, just that in crossing $t = 0$ their masses went to zero. Of course, these galaxies, planets, and stars emitted light, which travels along 45° lines in the space-time diagram of Figure 13-5. Should we be able to see these ancient objects that existed before $t = 0$?

In 1975, Sir Fred Hoyle published a paper entitled "On the Origin of the Microwave Background" in which he pointed out a familiar but often overlooked fact. In the process of scattering, the efficiency with which radiation is bounced around varies inversely as the mass of the particles that do the scattering. If the mass is low, the amount of scattering is high. In fact, at $t = 0$ the masses of particles is zero. Consequently, radiation interacting

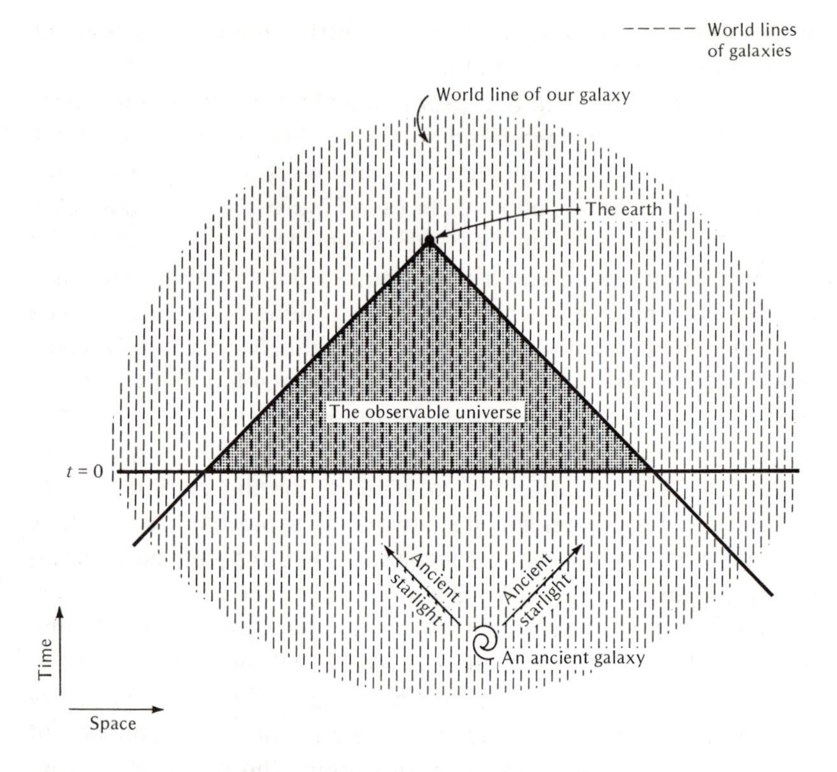

Figure 13-5. The whimper universe in space-time. Since the whimper universe is not expanding, the world lines of objects are parallel. Starlight from galaxies existing prior to $t = 0$ is completely thermalized as it crosses the $t = 0$ interface. This thermalized radiation is observed as the 3° blackbody background.

with zero mass particles at $t = 0$ will be completely scattered and thermalized. In the whimper cosmology, therefore, *the 3° blackbody background is simply the smoothed-out starlight from galaxies that existed prior to t = 0.* The microwave background, whose isotropy poses huge theoretical problems for the big bang astrophysicists, has a simple and straightforward explanation in a whimper universe.

In examining Figure 13-5, we notice that the world lines of particles extend to times earlier than $t = 0$. Material existing prior to $t = 0$ could, therefore, have an effect on the masses of objects existing in our universe today. Indeed, it might seem that the masses of objects existing today should be infinite because they can receive mass interactions from material that existed billions upon billions of years prior to $t = 0$. Obviously, this is an erroneous conclusion. Hoyle copes with this apparent paradox by proposing that the nature of the mass interaction changes upon crossing the $t = 0$ boundary. Material existing after $t = 0$ makes a *positive* contribution to the mass of objects, while material existing prior to $t = 0$ makes a *negative* contribution. In this way we are assured of the fact that the masses of people, trees, and

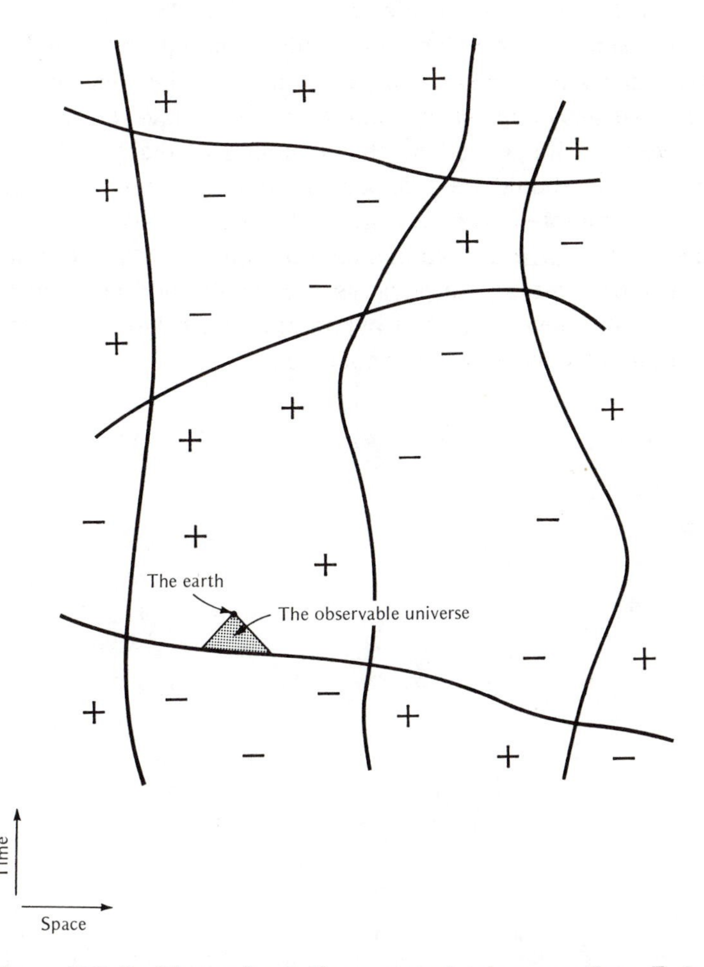

Figure 13-6. Positive and negative aggregates in space-time. To insure that the masses of particles are not infinite, perhaps all of space-time is divided into cells that alternately give positive or negative contributions in mass interactions.

planets around us today are not infinite. Indeed, all of space-time in a whimper cosmology may be divided up into regions of positive and negative aggregates, as shown in Figure 13-6. In looking back toward an interface between positive and negative aggregates, astronomers see phenomena that give the illusion of a big bang. Actually, however, their observations could always be explained by the fact that the masses of particles passing across such boundaries go to zero.

It should be emphasized that this new whimper cosmology is diametrically opposed to the mainstream of modern astronomy and astrophysics. Most scientists would contend that Hoyle's ideas are bizarre and fantastical. In fact, the whole business smells a little bit like astrology—a subject that would have us believe that our lives are influenced by the locations of distant objects in space. Yet, it should be emphasized that the whimper cosmology is in complete agreement with all of our observations and knowledge of the universe. We can only be grateful that there are competent astronomers who possess the genius and courage to propose radically new ideas and hypotheses from which we can construct a totally new view of the cosmos.

Epilogue

In these few pages we have attempted to present some basic ideas about the universe that have emerged over the past few years. In particular, we have placed special emphasis on theoretical research and how our discoveries might help in explaining our observations. One of the important questions that comes to mind is; How much of the material in this book is actually correct? Of course at the present time we have no way of knowing. In cases where detailed astronomical observations planned for the 1980s will prove enlightening, we have tried to point the way in which ideas might change. But nevertheless, we cannot help wondering what our theories will be like 50 or 100 years from now. We are reminded of some of the ancient ideas about cosmology, such as the Ptolemaic system, which was taken as a model of physical reality for over 1000 years. We realize today that the acceptance of the Ptolemaic system was based on the general philosophical orientation of ancient peoples along with a limited quantity of available relevant data. If you lived in Greece over 2000 years ago, you would probably prefer to think in terms of epicycles, even though people such as Aristarchus had suggested that all the planets revolve around the sun. We are quick to realize that any of our modern ideas must incorporate all available data and observations, but how has our psychological orientation affected the concepts put forth in this book? The general approach of all modern physical science is entirely mechanistic. It is mechanistic not in the crude sense of gears, levers, and pulleys, but rather in terms of trying to reduce all reality to concrete physical laws wherein the only truly important quantities are those that we can measure with machines such as spectrographs, galvanometers, and photographic film. In the past, people thought of the planets as gods and attributed specific spiritual qualities to what they saw in the sky. It is argued by modern scientists that such ideas contribute nothing of value to our understanding and, indeed, lead us astray. But when these same scientists walk up to their telescopes or cyclotrons, they carry with them a set of axioms that are just as much a part of their psyche as the unquestioned concepts in the mind of the ancient Babylonian astronomers climbing to the top of a ziggurat. It is obvious, therefore, that if we knew more about ourselves, we would be able to discover and understand more about the nature of the universe.

Glossary

absolute magnitude A measure of the real brightness of a star; the apparent magnitude a star would have at a distance of 10 parsecs.
absolute zero Theoretically, the lowest possible temperature; a temperature of $0°K = -273°C$.
aphelion The point in the orbit of a planet that is farthest from the sun.
apparent magnitude A measure of how bright a star appears to be in the sky.
astronomical unit A unit of distance defined as the semimajor axis of the earth's orbit; the average distance between the earth and the sun (93 million miles).
atom The smallest particle of a chemical element that still retains the properties characterizing that element.
barred spiral galaxy A spiral galaxy in which the spiral arms originate from the ends of a "bar" running through the center of the galaxy.
big bang theory A theory of cosmology in which the universe is supposed to have begun with a primeval explosion.
binary star A double star; two stars revolving about each other.
black hole An object having a gravitational field so intense that nothing, even light, can escape from it.
blackbody An ideal object that absorbs and reemits all radiation incident upon it.
calculus A branch of mathematics dealing with infinitesimal quantities.
cD galaxy A supergiant elliptical galaxy.
celestial mechanics A branch of astronomy dealing with the motions of the planets and other bodies in our solar system.
Cepheid variable A particular type of yellow supergiant pulsating stars.
classical mechanics A branch of physics dealing with the behavior of physical objects.
Clouds of Magellan Two nearby galaxies visible to the naked eye from southern latitudes.
cluster of galaxies A system of galaxies containing a few to thousands of galaxies.
compact galaxy A small galaxy having a high surface brightness.
conic section A curve that may be obtained by cutting a circular cone with a plane.
cosmological model A specific theory of the organization and evolution of the universe.
cosmological redshift The redshift associated with the general expansion of the universe.

cosmology A branch of astronomy dealing with the organization and evolution of the universe.
deferent A stationary circle along which an epicycle moves in Ptolemy's system.
Doppler effect (Doppler shift) The apparent change in the wavelength of radiation due to the relative motion between the source and the observer.
eclipsing binary star A binary star that is oriented to our line of sight in such a way that each star periodically passes in front of the other.
electromagnetic radiation Any type of radiation consisting of waves of oscillating electric and magnetic fields; includes visible light, radio, infrared, ultraviolet radiation, as well as x-rays and gamma rays.
electromagnetic spectrum The whole array of electromagnetic waves.
electromagnetic theory A branch of physics dealing with electric and magnetic fields as well as electromagnetic radiation.
electron A negatively charged subatomic particle usually found in orbit around the nucleus of an atom.
ellipse A conic section formed by cutting completely through a circular cone with a plane.
elliptical galaxy A galaxy having an elliptical shape but no spiral arms.
epicycle The circular orbit of a planet that moves along a deferent in Ptolemy's system.
event horizon A place beyond which it is impossible to communicate.
exit cone An imaginary cone used to distinguish between light rays that can and cannot escape from an intense gravitational field.
field equations A set of equations in general relativity that describes the curvature of space-time.
galaxy A large assemblage of stars, usually hundreds of billions of stars.
gamma ray A high-energy photon of electromagnetic radiation.
geodesic equations A set of equations in general relativity that describes the paths followed by particles and photons in curved space-time.
gravitational redshift The redshift caused by a gravitational field; the slowing-down of clocks in a gravitational field.
gravitational wave A changing gravitational field; oscillations in the curvature of space-time.
Hubble classification A scheme of classifying galaxies according to their appearance.
Hubble law A relation between the recessional velocity of remote galaxies and their distances.
hyperbola A conic section formed by cutting a circular cone with a plane.
infrared radiation Electromagnetic radiation having wavelengths slightly longer than the wavelength of red light visible to the human eye.
irregular galaxy An asymmetrical galaxy; neither an elliptical nor spiral galaxy.
Kepler's laws Three laws discovered by Kepler describing the motions of the planets.
Kerr solution An exact solution to the field equations describing a rotating black hole.
Kruskal solution A solution of the Einstein field equations that com-

pletely describes the geometry of space-time surrounding a static black hole.

light Electromagnetic radiation visible to the human eye.

light curve A graph that displays the variation in brightness of a star.

light year The distance light can travel in one year, approximately 6 trillion miles.

local group The cluster of galaxies to which our own galaxy belongs.

magnitude A measure of the brightness of an object such as a star or galaxy.

Messier catalog A catalog of nonstellar celestial objects compiled by Charles Messier in the eighteenth century.

N galaxy A galaxy with a starlike nucleus surrounded by a faint haze.

nebula A cloud of interstellar gas or dust.

negatively curved space A space in which "parallel" lines ultimately diverge.

neutron A subatomic particle having no electric charge and a mass very nearly equal to that of a proton.

neutron star A star composed almost entirely of neutrons.

New General Catalogue A catalog of star clusters, nebulae, and galaxies compiled by J. L. E. Dreyer in the late nineteenth century.

Newton's laws The laws of mechanics and gravitation formulated by Sir Isaac Newton.

nova A star that has a sudden outburst of radiant energy.

nucleus (of an atom) The heaviest part of an atom consisting mostly of protons and neutrons.

nucleus (of a galaxy) The central condensation of stars and gas at the center of a galaxy.

orbit The path of a body that is revolving about another body or point.

parabola A conic section formed by cutting a circular cone with a plane.

parallel transport (of vectors) A specific method of moving a vector from one location to another.

parsec A measure of distance (1 pc = 3.26 light years).

Penrose diagram A map of all space-time associated with a black hole.

perihelion The place in the orbit of an object revolving about the sun at which the object is closest to the sun.

photon A discrete unit of electromagnetic energy.

photon circle (photon sphere) An unstable circular orbit for light about a black hole.

positively curved space A space in which "parallel" lines ultimately converge.

primeval fireball The exertmely hot gas that is presumed to comprise the entire mass of the universe at the time of the "big bang."

prism A wedge-shaped piece of glass that can be used to disperse white light into the colors of the rainbow.

proton A subatomic particle with positive electric charge; one of the two primary constituents of the atomic nucleus.

pulsar A rapidly pulsating radio source.

quasar A starlike object having a very large redshift.

radio astronomy The branch of astronomy dealing with radio waves from space.

radio telescope A telescope designed to detect radio waves.

red giant A large, cool, red star having a high luminosity.

redshift A shift in wavelength toward the red end of the spectrum.
Schwarzschild solution An exact solution to the field equations discovered by K. Schwarzschild.
semimajor axis One-half the longest distance across an ellipse.
Seyfert galaxy A spiral galaxy (usually a radio source) having a bright nucleus of the type first described by C. Seyfert.
singularity A place where the intensity of the gravitational field is infinite.
solar system The system consisting of the sun, the planets, their satellites, asteroids, comets, and so on.
spectral lines Lines (either dark or bright) that appear in a spectrum.
spectrum The rainbow of colors obtained by passing a beam of white light through a prism.
spiral galaxy A rotating galaxy having pinwheellike arms of gases and young stars winding out from its nucleus.
star A self-luminous sphere of gas.
steady state theory A theory of cosmology that assumes that the overall properties of the universe do not change with time or location.
stellar evolution The changes that take place in the properties of stars as they age.
tensor A certain type of mathematical quantity.
thermonuclear energy The energy that can be released through the thermonuclear reactions.
thermonuclear reaction A nuclear reaction that results from the collisions of nuclear particles at high speeds.
Third Cambridge Catalogue A catalog of radio sources.
ultraviolet radiation Electromagnetic radiation having wavelengths slightly shorter than violet light visible to the naked eye.
universe The totality of all matter, radiation, space, and time.
vector A quantity that has both magnitude and direction.
whimper theory A theory of cosmology in which the universe is not expanding and the masses of particles change with time.
white dwarf A small dense star; one possible end-point of stellar evolution.
white hole The time reversal of a black hole; place where matter and energy gush up.
wormhole A connection in space-time between two universes or two distant parts of the same universe.
x-ray A type of electromagnetic radiation having wavelengths between ultraviolet radiation and gamma rays.
ziggurat A temple and astronomical observatory in ancient Babylonia.